"Once again, Tom Wheeler makes sense out of the dizzying technological changes that often seem to initially befuddle and beset us before they come into sharper focus, a focus he brings to each page and each new idea. Wheeler understands in his bones that 'what is past is prologue,' and so he correctly anchors the new in the context of what has taken place before. Ecclesiastes is always right: 'There's nothing new under the sun,' but it sometimes takes an original thinker to make clearer the 'mess' in front of us."—**Ken Burns**

"In this tour de force, Wheeler not only unpacks the challenges that the new gilded age poses to consumer privacy, competition, truth, and trust but also highlights ways to safeguard them and us. An eye-opening guide to a more hopeful future!"—**Kathleen Hall Jamieson**, director, Annenberg Public Policy Center, University of Pennsylvania

"Wheeler is one of the major global players on Technology and Media regulation in the twenty-first century. *Techlash* is a powerful book that speaks to some of the most important issues facing our society. Wheeler's expertise as both a business leader and America's top media regulator offers a unique and trenchant perspective that makes this a must read for anyone concerned about technology's impact on our lives . . . and on our children's lives."—**James P. Steyer**, CEO and founder of Common Sense Media, Stanford professor, and bestselling author

"The clarion call by Tom Wheeler for a new model of governance in the internet age demands our attention. Wheeler's thoughtful case for agile oversight is grounded in history and should be read by all who care about public policy."—**Phil Weiser**, Attorney General of Colorado

"Wheeler brings an invaluable mix of insight, experience, and historical knowledge to this critical challenge for our age: how do we protect innovation and the fruits of the digital revolution while also protecting individual rights and the broader public interest? At a moment when 'technology is policy' and the coders hold the keys, the need for wise and nimble regulation grows more apparent by the day. Wheeler is uniquely positioned to sort through the challenges, choices, and competing values, and this book will be essential reading for all those invested in platform accountability and the health of our information ecosystem."—**Nancy Gibbs**, director, Shorenstein Center for Media and Policy, Harvard Kennedy School; former managing editor *TIME*

"Wheeler's *Techlash* is an urgent and timely work of public service. For too long, the American people have been left to defend themselves against powerful tech companies that erode their privacy, addict their kids, and undermine our democracy. Wheeler's lucid and historically grounded book describes the utter inadequacy of our existing regulatory structure to defend the American people against technologies moving at the speed of light, and he makes a compelling case to stand up a new federal body to oversee digital platforms and defend the public interest—just as we did for radio, air travel, and pharmaceuticals at previous moments in our history. If we choose to meet our moment, *Techlash* shows us the way."—**Senator Michael Bennet**

"In graceful and concise language, Tom Wheeler brings his entrepreneurial and regulatory experience to explain and demystify the impact of digital technology on our economy and society and how government must come off the sidelines to protect the public interest. Wheeler calls for government oversight with a new, flexible regulatory framework fit for the speed of technology that would protect the public while encouraging innovation. *Techlash* is an outstanding and necessary read for all who want to understand the impact of our digital economy and how to curb its excesses without curbing its benefits."—**Senator Peter Welch (D-VT)**

"Brilliant! Every member of Congress and every state AG needs to read this book now." —**Roger McNamee,** author of the *New York Times* bestseller, *Zucked: Waking Up to the Facebook Catastrophe*

TECHLASH

Who Makes the Rules in the Digital Gilded Age?

TOM WHEELER

BROOKINGS INSTITUTION PRESS
Washington, D.C.

Published by Brookings Institution Press
1775 Massachusetts Avenue, NW
Washington, DC 20036
www.brookings.edu/bipress

Co-published by Rowman & Littlefield
An imprint of The Rowman & Littlefield Publishing Group, Inc.
4501 Forbes Boulevard, Suite 200, Lanham, Maryland 20706
www.rowman.com

86-90 Paul Street, London EC2A 4NE

Distributed by NATIONAL BOOK NETWORK

The Brookings Institution is a nonprofit organization devoted to research, education, and publication on important issues of domestic and foreign policy. Its principal purpose is to bring the highest quality independent research and analysis to bear on current and emerging policy problems.

British Library Cataloguing in Publication Information Available

Library of Congress Cataloging-in-Publication Data Available

ISBN 978-0-8157-3993-7 (cloth)
ISBN 978-0-8157-3994-4 (electronic)

Library of Congress Control Number: 2023944517

∞™ The paper used in this publication meets the minimum requirements of American National Standard for Information Sciences—Permanence of Paper for Printed Library Materials, ANSI/NISO Z39.48-1992

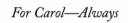

For Carol—Always

Contents

Contents

PREFACE

It is "an old pattern in American economic history," historian John Steele Gordon explained. "Whenever a major new force—whether a product, technology, or organizational form—enters the economic arena, two things happen. First, enormous fortunes are created by entrepreneurs who successfully exploit the new, largely unregulated economic niches that have opened up. Second, the effects of the new force run up against the public interest and the rights of others."[1]

This "old pattern" echoes today. New digital technology has changed commerce and culture, creating great wealth in the process, all while being essentially unsupervised. The lack of consideration for the consequences of the actions taken by these new enterprises harkens back to the industrial revolution and the Gilded Age it created. As technological advancement has continued, bringing new realities such as the metaverse and artificial intelligence, the history of innovactors acting without oversight of the consequences continues.

While history does not precisely repeat itself, there are common themes that connect the challenges—and solutions—of the twenty-first century with their Gilded Age antecedents. This is not the first time society has been confronted by technology-enabled barons of marketplace power. Similarly, it is not the first search for solutions to protect "the public interest and the rights of others."

It is, however, a legacy moment for this generation to determine whether, and how, it will assert the public interest in the new digital environment. The torid expansion of artificial intelligence only increases the importance of such legacy-making decisions.

In the original Gilded Age, government ultimately ceased being a spectator and became a participant in the new economy. Such participation required thinking anew about how to address the challenges presented by the industrial economy. The solutions that were developed—antitrust law and regulatory oversight—were unprecedented. The new Gilded Age requires a similar dedication to thinking anew.

Thus far in the digital era we have delegated the establishment of economic and social norms to those who can write software code. It is an admirable ability, but it is not a qualification to write the rules that increasingly dictate individual lives and affect the operation of American democracy.

As happened in the Gilded Age, the time has come to reassert the public interest. Just as today's innovators looked beyond historical norms to develop new capabilities, so must the response to what they have imposed look beyond the regulatory oversight models of the past. The solutions that will create a fair and sustainable digital economy cannot simply be retreads of industrial era ideas. Today is already vastly different from yesterday, and tomorrow promises even more change. In preparation for that new reality, our decisions must be as bold and innovative as the new technology itself.

It is time to make history again.

***

Samuel Clemens—Mark Twain—created the Gilded Age moniker. *The Gilded Age: A Tale of Today* was a novel he co-wrote to satirize the economic excesses, personal greed, and political corruption of the late nineteenth century.[2]

In Twain's telling, the era was not a "golden" age—something pure and solid like a bar of gold—rather, it was a "gilded" age. The definition of "gild" is to cover something of lesser value with a coat of gold to make it look like what it is not.

During the original Gilded Age, the flood of technology-based mechanical innovations and the wealth they created put a gold-hewed gloss on the darker effects that accompanied those developments. As we enter the second quarter of the twenty-first century and a new Gilded

Age, once again wonderous and beneficial new products and services are reshaping how we work and live. Yet . . .

Once again, these innovations have created great disparity in the distribution of economic opportunity and wealth.

Once again, the innovators are making the rules for the new marketplace.

Once again, a shiny patina hides the darker realities.

Once again, it is time to act.

The rampant industrial capitalism of the original Gilded Age embraced new technologies with little consideration of the consequences. Such practices were ultimately brought in check by government intervention to protect consumers, workers, and the competitive marketplace. We have yet to see a similarly meaningful response to the darker aspects of rampant internet capitalism.

The record will attest that I am a capital C Capitalist. The historical record makes it clear that capitalism is the most efficient way to allocate resources. It is also true that capitalism flourishes when there are guardrails in place to protect against its inherent incentive to excess. Even such a free marketeer as Adam Smith recognized that for market forces to exist there had to be basic behavioral expectations.

The rules put in place to oversee the industrial revolution—protecting consumers, workers, and the competitive market—also ended up protecting capitalism. We forget how communism on the left and fascism on the right were popular "solutions" proposed to deal with the insecurities of the industrial era. It is a history we should remember as today authoritarians and the simplicity they promise are on the rise, in part because of the uncertainty created by the new digital era. The installation of basic behavioral rules for internet capitalism would be one of the most pro-capitalism actions that could be taken at this time.

The oversight of industrial capitalism required a departure from the philosophies that had governed its predecessor, agrarian mercantilism. The laissez-faire concepts that had worked for an economy built around individual farmers and small localized businesses were inadequate for industrial mass production enabled by the new rail and telegraph networks and abetted by new mechanical equipment. In a similar manner, oversight of internet capitalism built around high-speed networks, digital assets, powerful computing, and network effects necessitates the development of new regulatory concepts different from those engineered for industrial capitalism.

***

Digital technology has created new products, new services, and new wealth with little consideration of the consequences and with little oversight to protect the public interest. Use of the new technology has created some of the most powerful and influential companies in the world.[3] Yet, despite their influence and the promises the companies continually make about solving the problems they create, technology companies are increasingly less trusted.[4]

Once-small startups led by creative entrepreneurs have exploited internet-based technology and its lack of oversight to become giant corporations. In the process they were permitted to ride roughshod over accepted norms and behaviors. It is not that the companies and their leaders are bad actors, just that they have been given free rein over their behavior and have taken advantage of that lack of supervision to act in their own economic self-interest. While no one expects them to act like the Red Cross, they too often have behaved as though they are "too big to care." The challenge we now face is whether they have become "too big to fix."[5]

The internet started out with the hope that its distribution of economic activity would be the great democratizer by removing the barriers to everything from the flow of news to de-monopolizing local taxi service. The industrial economy was built on networks that performed their tasks at centralized physical hubs (think railroad switching yards) around which congregated centralized production. The physical structure of the internet, in contrast, is a collection of disparate interconnected networks whose functions are distributed across a fishnet-like architecture where each knot of the net performs the sorting and routing function of a rail yard.

It was thought the internet's distributed architecture would similarly distribute power away from centralized institutions. In many cases this has been true as new distributed businesses such as e-commerce successfully disintermediated traditionally centralized enterprises. At the same time, however, the internet has created new centralized powers. These new powerhouses built a closed superstructure on top of the internet. That superstructure—private computer banks linked to the internet's low-cost connections—became a platform that re-centralized network-based economic activity. The companies that built the platform superstructure—firms like Google (Alphabet), Facebook (Meta), Amazon, Apple, and Microsoft—have thus become systemically important "platform companies," a collective term used throughout this book.

The combination of low-cost computing with low-cost connectivity is the economic engine of the twenty-first century. Everything from automobiles to elevators contains microcomputers that report their status to central databases. Tracking the condition of a car's brakes or monitoring the machinery of an elevator goes on silently in the background of our lives producing improved safety and productivity.

The platform companies have applied similar techniques to consumer-facing services. In these applications, however, the information being collected is not relatively anodyne mechanical performance data, but heretofore private information. Having siphoned such personal information from consumers, the platform companies convert what had once been private into a corporate asset.

It is a brilliant business model. The internet provides a low-cost means of collecting personal information; then that data is locked away to be unavailable to others, including potential competitors. The resulting data bottleneck can only be unlocked by paying the platform what the market will bear. The activity that generates the highest return on such data is the targeting of advertising or other transactions based on knowledge of the personal proclivities of each consumer.

The consequences of the platforms' business model not only include the invasion of personal privacy and the control of that personal information to thwart competition and innovation, but also the disassembly of a common collection of facts. In the process, the hope that the internet would become the great disintermediator of economic activity and democratizer of meaningful expression and dialog has been dashed. In place of such ideals, corporate bottlenecks have been created to centralize economic activity, control the flow of information, and manipulate users.

To meet such challenges is as American as apple pie. The story of the late nineteenth and early twentieth centuries is the tale of confronting and overcoming the effects of technology-driven change. If American society was once able to overcome the industrial barons' abuse of marketplace competition, consumer safety, and workers' rights, we should believe it is possible to accomplish similar goals once again.

It was a decades-long fight to bring the corporate power of the original Gilded Age into a semblance of harmony with the public interest. We face a similar challenge in the Gilded Age of the twenty-first century. The short attention span that characterizes our time too often prevents us from seeing this historical analogy or appreciating its lesson that solutions will not come overnight. But come they must.

***

From 2013 until 2017, I had the privilege of a unique perspective on both the technological changes shaping our future as well as society's attempts to deal with those shifts. As chairman of the Federal Communications Commission (FCC), with responsibility for the broadcast, cable, telephone, and internet networks, I saw firsthand how the churning of technology had delivered a previously unimaginable cornucopia of products and services. I also witnessed how old industrial-era statutes, coupled with a lack of political will and technical understanding, allowed the barons of the internet to behave in a manner like the industrial barons of history to become pseudo-governments to make their own rules. This book is an effort to reflect the learnings from that experience, put them in historical perspective, and translate them into a policy framework for tomorrow.

History is the story of our past that shapes the present and informs the future. For the purposes of this book, that history begins at the end of the American Civil War as the industrial revolution morphed into what would become the Gilded Age that would last until the early twentieth century. What follows is a collection of stories about the industrial revolution, how similar stories play out today in the introduction of new technology, and what these experiences tell us about dealing with our digital future.

This book is divided into five parts, as described below.

PART I: OUR MOMENT IN HISTORY

Chapter 1, "Echoes of the Gilded Age"—Both the original Gilded Age and the new era were made possible by government investments in transformational technologies. The railroad and telegraph of the nineteenth century, as well as the internet, microchips, and even the Google algorithm were all assisted by public policy and the expenditure of public assets.

When a handful of industrial barons seized upon their era's new capabilities to create new products and services, the result was similar to the effects coming from a handful of internet barons today: an acceleration of the pace of life, a disparity in wealth, the destruction of small businesses, creation of monopolies, consumer harm, and even fake news.

The late nineteenth and early twentieth centuries witnessed an epic struggle as corporate interests fought legislation and regulation to assert the public interest. We are today presented with the same question that arose in the industrial Gilded Age: Will there be rules for the new economy, and who will make those rules?

Chapter 2, "This Is NOT the 'Fourth Industrial Revolution'"—To many observers, including the World Economic Forum, today's economy is an extension of the industrial revolution that drove the Gilded Age—it is the "Fourth Industrial Revolution."[6] The causes and effects of the digital revolution, however, are different from the industrial revolution. As a result, protecting the public interest in the digital era cannot only rely on industrial era assumptions about market behavior.

While some of the forces that drove the industrial revolution remain—such as the economics of scope and scale—new forces have also taken hold to shape the new era. The industrial economy was a pipeline economy built on hard assets. The internet economy is a platform economy built on the soft assets of computer code. Not only do the assets behave differently, but also the process of their monetization is different. Industrial activity is built on a linear production model. In contrast, the activity of digital platforms is an exponential two-sided process in which the platform is a mechanism for pairing digital assets.[7]

Our response to the digital revolution is certainly informed by the experiences of the industrial revolution. However, the search for solutions to the new digital challenges begins with an understanding of the difference between the digital economy and the industrial economy. Those differences mean that replicating what worked yesterday is inadequate for today and what is ahead.

Chapter 3, "Closing the Open Internet"—The internet that underpins and enables the digital platforms is designed to be open and interconnected. The dominant digital platforms have perverted that openness to create a closed superstructure that exploits personal privacy, destroys marketplace competition, and disseminates untruths.

The platforms have created a self-perpetuating digital chain reaction. The seminal force in this business model is how the companies confiscate personal information and convert it into their own corporate asset. Once in control of the asset, the companies then become its gatekeeper to deny access to potential competitors and others. Then the platforms combine that market control and data control to manipulate users through control over the selection of the news and information consumers receive. How users respond to that process then creates additional data about each user—and the chain reaction begins again.

It was hoped that the internet would become the impetus for the distribution of economic activities and power. Instead the open network has created closed platforms and a new collection of centralized corporate powerhouses.

PART II: YOU AIN'T SEEN NOTHIN' YET!
Chapter 4, "The Metaverse"—The next generation of online platforms is the so-called metaverse. As Mark Zuckerberg explained just before changing his company's name from Facebook to Meta Platforms, "you can think about the metaverse as an embodied internet, where instead of just viewing content—you are in it."[8] Using virtual reality (VR), augmented reality (AR), artificial intelligence (AI), and constant connectivity, the metaverse transforms the online experience from 2-D observation

to 3-D participation. It is the transition from social media to social virtual reality.

The metaverse brings exciting new opportunities. In education, for instance, students can be transported to witness historical moments, and doctors trained on virtual patients. This next generation of technology, however, has descended on us before we have even dealt with the harmful realities of today's internet. The metaverse can only expand those harmful effects.

Because metaverse avatars will be personally identifiable, it will be necessary to collect even more personal information about users. Today, the platforms track our clicks, and likes; the metaverse will track our bodily functions. The headset (and soon glasses) that transport users to the metaverse also takes from those users' biometric data such as eye movements, heart rate, and facial expressions. The result is to invade each user's psychological makeup more deeply than anything seen to date. That the collection of such personal information is being carried out by some of the same companies that have already trampled on personal privacy in order to use the data to dominant markets and target misinformation should not go unnoticed.

Chapter 5, "Artificial Intelligence"—The Industrial Revolution that drove the Gilded Age was built on replacing and/or augmenting the physical power of humans. Artificial intelligence is about replacing and/or augmenting humans' cognitive power. Russian President Vladimir Putin observed that whoever leads in AI "will be the ruler of the world."[9]

After years of speculation and promise inside computer labs, AI broke loose in November 2022 with the launch of ChatGPT. The new AI software utilized a chatbot model that enabled a user to "talk" with the program using natural language. It was called "generative AI" because it generates answers based on the data it has been trained on.

Like the metaverse, AI can bring with it marvelous new capabilities. It can equip doctors with better analytical and prescriptive tools. Its ability to recognize objects can help individuals who are blind to see. It can increase business efficiency and reduce costs. It can be a virtual tutor for online learning.

At the same time, AI could accelerate the anti-consumer, anti-competition, anti-truth behaviors we are seeing in today's online services. AI will threaten white collar jobs and increase cybersecurity risks. Its use of data poses another threat to personal privacy and the potential for massive surveillance. And then there are the apocalyptic visions of AI defying its human creators.

Like the metaverse, the AI explosion is happening without any supervision to protect the public interest. AI technology is being released with fewer safety standards and less oversight than that required for a toaster. The difficulties we associate with today's online platforms resulted when they entangled themselves in our daily activities, amassed great wealth and political power, and were able to resist calls for public interest oversight. The lesson of the digital era thus far is a message about what happens when policymakers fail to get in front of the new technology's effects. That story cannot be allowed to repeat in a world driven by artificial intelligence.

PART III: WHO MAKES THE RULES?

Chapter 6, "When Innovators Make the Rules"—The financier of much of the Gilded Age, J. P. Morgan, defined the rules of the period as "a certain number of men who own property can do what they like with it."[10] In the digital age, Mark Zuckerberg's famous mantra "Move Fast and Break Things" expresses a similar conceit. The "things" to which Zuckerberg refers are not physical objects, but behavioral expectations that for years had been relied upon to provide economic and social stability. Moving "fast" is necessary to implanting such practices before the masses understand what is happening.

The digital revolution has thus created a conundrum. The vision and tenacity to break the rules has historically been the pathway to advancements in science, business, and the arts. The inventiveness of innovators that allows them to soar past the rules is cause for celebration as well as concern for its effects. In both the original Gilded Age and today, those effects often end up benefiting their creator more than the public that lives under the new order.

As a result of the excesses of the original Gilded Age, the government stepped in to balance risks and benefits. Currently, however, the lack of governmental leadership has empowered the digital platforms to assume the role of a pseudo-government to establish their own behavioral standards. The challenge presented by the rule breaking of the digital innovators is to balance their creativity and new products against how those activities affect the rights of others and the public interest.

Chapter 7, "The World's Greatest Business Model"—The barons of the Gilded Age built corporate behemoths. Rockefeller's oil, Carnegie's steel, and Vanderbilt's railroads were among the most valuable enterprises of their time. Today, Apple, Microsoft, Amazon, Alphabet/Google, and Facebook/Meta are among the most valuable companies in the world.

The riches of the digital giants have been made possible by the world's greatest business model: swipe and sell (or more precisely, rent) the capital asset of the twenty-first century—data.

Using the low-cost efficiencies of computing power and the low-cost delivery of the internet, the dominant digital platforms can collect and manipulate data on users at marginal costs that approach zero. Then, on the other side of the business model, the platforms use their dominant control of the information thus collected to charge an optimized price for its use by advertisers and others.

Chapter 8, "Where Is the Watchdog?"—The Gilded Age brought forth American leadership in implementing new oversight of the effects of the industrial economy. The laissez-faire policies that had governed agrarian mercantilism were exchanged for policies attuned to industrial capitalism. Today, meaningful American leadership is absent in the establishment of policies on a scale attuned to the new digital reality. Thus, not only do the companies make the rules, but so do our international allies and adversaries that have rushed to fill the void created by American inaction.

Internet connections that leap borders have allowed the dominant digital platforms to impose their own set of rules across the interconnected world. Beyond the United States, however, the interconnected world is pushing back. The result is a race among nations to establish de jure rules for online activity within their borders. Because of the seamless interconnection of the internet, these national decisions can become de facto rules across the globe, including in the United States.

The European Union has been first out of the gate with its Digital Markets Act (DMA) to focus on market behaviors, Digital Services Act (DSA) that focuses on content-related behavior, and the Artificial Intelligence Act (AIA) dealing with computer intelligence. The United Kingdom, after exiting the EU, is developing its own set of market-related and content-related policies. Fastest across the finish line has been the Chinese government with its authoritarian ability to dictate policy.

Compared with the EU, UK, and China, the United States is on the sidelines watching, the race is on to develop national policies that will be spread internationally by the same network that created the need for regulation in the first place.

PART IV: REASSERTING THE PUBLIC INTEREST

Chapter 9, "Designing Behavioral Expectations"—Technology, economic activity, and human behavior have changed. Regulation has not.

Accomplishing meaningful oversight necessitates reimagining the role of government away from the rigid and sclerotic practices of the industrial era to embrace the dynamism of the digital era. The digital companies broke the old way of doing things in the marketplace. Oversight in the digital era must similarly break with the old regulatory approach to embrace a new regulatory paradigm that reflects the new digital realities.

The companies of the digital era long ago discarded rigid top-down industrial management techniques in favor of agile processes capable of responding to the pace of change. Government has not kept pace to similarly evolve. As part of the new regulatory model, government structures must now catch up by developing a new approach to oversight that protects consumers and competition while also encouraging innovation.

Such oversight should focus on identifying and mitigating significant risks rather than dictating specific operational behavior. One way of accomplishing this is to use the standards-setting techniques utilized by digital companies to define how technologies function for the development of enforceable behavioral standards. Such standards are agile enough to keep up with changes in both technological capabilities and the marketplace. Bringing such a process to the establishment of behavioral standards would transform regulatory oversight from a micro-management model to one that focuses on risk management through enforceable behavioral codes.

Chapter 10, "Privacy by Design"—In the Gilded Age the extraction of industrial assets from the earth created environmental wastelands. The extraction of today's digital assets also carries an environmental cost—this time, however, it is the human environment. Mining for minerals has been replaced by mining for personal information. The digital platform business has been built on what Harvard's Shoshana Zuboff has labeled "surveillance capitalism."[11]

When asked, "What is Google?" its co-developer, Larry Page, responded, "If we had a category, it would be personal information . . . Your whole life will be searchable."[12] When he was Google CEO, Eric Schmidt explained we need not worry about such practices. "There is what I call the creepy line," he explained. "The Google policy is to get right up to that creepy line and not cross it."[13] Feel better?

The so-called privacy protections that the companies have created are actually permissions to exploit our personal information. Rather than pruning the ever-growing thicket of privacy invasions it is time for a wholesale replanting to establish an enforceable code to design for privacy.

Chapter 11, "Competition by Design"—The absence of effective competition has inoculated the dominant digital companies from focusing on what is best for customers versus what is best for the company.

Numerous lawsuits by both the federal government and coalitions of states seek to use the antitrust statutes developed in the Gilded Age to respond to the practices of the digital giants. Antitrust enforcement is important to pursue, but it is slow and even if successful cannot always reach noneconomic issues such as privacy or misinformation. As Congress considers how the antitrust statutes should be updated for the digital era, it is important to also pursue how the use of regulation can promote competition, not as an alternative to antitrust but as an addition.

The lynchpin of effective competition in the digital era is the interconnected nature of the internet itself. The dominant platform companies have achieved their position by embracing that interconnection when it serves their purpose (such as going to other sites to collect personal information) while destroying such interoperability when it would serve the purposes of their users and potential competitors (such as being able to connect with a friend using another platform).

The time has come for an enforceable code that encourages competition by opening platform data boards so that competition is based on better products, not just bigger data caches.

Chapter 12, "Truth and Trust by Design"—The Gilded Age was an era in which fake news flourished. The incentive to disregard facts if it increases revenue is unchanged in the twenty-first century. What has changed today is the unprecedented power of digital platforms to distribute such questionable material not only in a targeted, but also in a secret manner. It is a power built on software algorithms that select and target information based on corporate economic considerations rather than judgment about the information's veracity.

The activities of social media platforms have produced a technology-driven tribalism that is further enhanced by artificial intelligence and acted out in artificial reality. By distributing information, even of questionable veracity, to specific target audiences, the platforms are gnawing away at the shared knowledge that is essential for a democracy to function.

Editorial responsibility finally came to the Gilded Age's profits-by-untruth media as a matter of conscience. Newspaper editors banded together to introduce professional standards.[14] Today's algorithms, however, are soulless software without a conscience. The twenty-first century challenge becomes how society, acting through government, mounts such a demand for accountability while respecting the protections of the First Amendment to the United States Constitution.

The Constitution makes the American government's involvement a complex and fragile activity. Nonetheless, because editorial processes that preserve free speech and open information are a key component of democracy it demands we make the effort.

PART V: CONSEQUENCES WE CONTROL

Chapter 13, "Time to Make History Again" — This time around it is the internet barons who are setting the rules and thus far have prevented the people's representatives from establishing digital public interest protections.

In the original Gilded Age, the industrial barons set the rules, up until the point when the representatives of We the People stepped in to balance the scales with the broader public interest. The result was not only new protection of consumers, workers, and the competitive market, but also the preservation of industrial capitalism. We are now at a similar moment regarding rebalancing the scales of internet capitalism.

In one sense, this is a question of whether autocracies or participatory democracies determine the rules for the digital economy. Autocratic digital platform executives moved early on to dictate the rules governing their operations. Autocratic governments are similarly moving with dispatch to shape rules to their advantage both domestically and internationally.

The failure of the United States to develop meaningful digital policies does not mean that there will be no regulation, but that the rules will be made by others. The fact that the old American regulatory model is struggling in these new times is not an argument against regulation, but rather a demonstration of the need for a new regulatory paradigm to deal with the new digital realities.

There will be many debates regarding the best way to address the challenges created by the new Gilded Age. As these come forward, it is essential that we recall the overriding lesson of history: what made America great was confronting, not fleeing, its challenges. We must rekindle that attitude and restart the process of American policy leadership. Any such effort must begin with asking the question, "Who makes the rules in the digital Gilded Age?"

Part I

Our Moment in History

"The future is unknowable, but the past should give us hope."
—Winston Churchill,
A History of the English Speaking Peoples

CHAPTER 1

Echoes of the Gilded Age

EARLY SPRING WEATHER IN THE NATION'S CAPITAL CAN BE A CONTEST between snow and daffodils. On March 4, 1905, the patches of snow from the night before were disintegrating under a bright sun as Theodore Roosevelt stood before the Capitol to take the presidential oath of office. His message to the economic powers of the day was less sunny than the weather.

It was late in the Gilded Age; the nation was awash in the benefits of innovative new technologies and economic prosperity. Per capita gross domestic product (GDP) was soaring (it would end up multiplying more than seven times from 1865 to 1920).[1] The percentage of Americans living in the nation's cities had doubled to more than 40 percent,[2] and the nation's wealth had quadrupled.[3] The average American could expect to live to age fifty,[4] thanks in large part to industrial-scale improvements in both sanitation and health care.[5] Vanderbilt's railroads, Carnegie's steel, and Rockefeller's oil—all economically dominant and politically powerful—were reshaping commercial activity and improving daily living.

Amid the glitz and glitter, it was also a time of growing economic inequality, abuse of consumers and workers, and corporate subversion of the people's democratic institutions.

In his inaugural remarks, President Roosevelt discussed this national dichotomy. While the industrial economy had produced "marvelous material well-being," it also generated "care and anxiety" that were "inseparable from the accumulation of great wealth."[6]

3

Speaking for the average citizen buffeted by technology-driven, market-concentrating change, Roosevelt observed, "Modern life is both complex and intense, and the tremendous changes wrought by the extraordinary industrial development of the last half century are felt in every fiber of our social and political being."[7]

It was a message that could have been delivered today by simply replacing the word "industrial" with "digital." The extraordinary evolution from an analog economy built on machinery, to a digital economy built on algorithms has created a new Gilded Age that is also "felt in every fiber of our social and political being."[8]

Roosevelt concluded with a message that should be delivered today. The time has come for the nation to "approach these problems with the unbending, unflinching purpose to solve them aright."[9]

THE GILDED AGES

Both the original Gilded Age and that of the twenty-first century were enabled and sustained by the introduction of new networks. In the early iteration those networks were the steam railroad, the world's first high-speed network; followed shortly by the telegraph, the first electronic network.

Prior to the arrival of the steam locomotive midway through the first half of the nineteenth century, geography had ruled. Distance and topography determined the scope of economic activities. The consequences of the steam railroad, however, were more than the harnessing of mechanical muscle. The railroad was, as historian Jacques Barzun observed, "The completest change in human experience since the nomadic tribes became rooted in one spot to grow grain and raise cattle."[10]

Steam rolling on steel meant the agrarian revolution yielded to the industrial revolution. Because early locomotives indefatigably moved five to ten times faster than animal power, they vastly expanded the serviceable area of economic activity. Because heavy and bulky raw materials could now be transported economically, production could be centralized in order to enjoy scope and scale economies. Running in the opposite direction, the railroad then created an interconnected mass market. These activities were controlled through sparks on the telegraph wire. The result was the manufacture and delivery of an abundance of useful products at low prices.

The Gilded Age of the twenty-first century is built around the descendants of those earlier networks. The computing power of microchips has replaced the mechanical power of giant steam engines to speed content across a new network. The binary dot-dash of messages over telegraph wires has evolved to become the distribution of binary zeros and ones over wires, fiber, and through the air. The internet and digital algorithms have upended the industrial model.

Centralized economic activity in which Detroit produced vehicles, Pittsburgh produced steel, and New York produced news yielded as software algorithms replaced industrial machinery, digital information replaced hard industrial assets, and all were connected by a ubiquitous network. The new Gilded Age is constructed on a symbiotic relationship between software algorithms and internet connectivity. Like their nineteenth-century predecessors these have redefined both commercial enterprises and cultural behavior. "Software is eating the world,"[11] Marc Andreessen, developer of an early web browser, famously observed.

Digital Echoes

History has recorded that while the industrial revolution delivered new products at lower prices, it also produced social and economic upheaval. Digital technologies are now writing a new history with a similar storyline: the production not only of new products and services, but also new problems.

Income Inequality

Not since the Gilded Age has "America witnessed a similar widening of the income gap," the *Economist* observed.[12] At the height of the first Gilded Age, the top decile of Americans commanded more than 45 percent of the nation's gross income. Today, the top decile receives more than 50 percent of all income.[13] "The U.S. has the highest level of economic inequality among developed countries," Nobel economics laureate Joseph Stiglitz wrote in 2018.[14]

The economic expansion of both eras showered a small group with staggering wealth. Gilded Age tycoon John D. Rockefeller is reputed to have been the wealthiest individual in American history, with a net worth of between $300 and $400 billion in today's dollars.[15] Vast personal fortunes were created by, among others, steel baron Andrew Carnegie, railroad magnate Cornelius Vanderbilt, and financier J. P. Morgan.

In 2020, digital baron Jeff Bezos was reportedly the richest person in the world, with a net worth of over $212 billion. Microsoft founder Bill Gates ranked fourth, Facebook's Mark Zuckerberg fifth, and Google founders Larry Page and Sergey Brin ranked sixth and seventh respectively.[16] While subsequent economic difficulties took their toll to reshuffle the rankings, they all remain among the world's richest individuals. President Biden reported to Congress that during the COVID-19 pandemic—a period when 20 million Americans lost their jobs—roughly 650 billionaires saw their net worth collectively rise by more than $1 trillion.[17]

Market Concentration

It was John D. Rockefeller's lawyer, Samuel Dodd, who in 1882 came up with the idea of using a legal fiction—a trust—as a tool to eliminate competition.[18] A trust was not a productive corporation; rather it was a legal means to force cooperation through ownership. Shareholders of competitive companies would turn their stock over to trustees who, in turn, would pay them dividends inflated by the resulting noncompetitive market. The trust Rockefeller ultimately created—the Standard Oil Trust—controlled 91 percent of oil production and 85 percent of all final U.S. sales. This dominance was distributed across sixty-seven subsidiaries in a gigantic, coordinated concentration of power and market control.[19]

In the new Gilded Age, Google controls over 90 percent of online search worldwide[20] and 56 percent of domestic search advertising.[21] While not a trust, Google's management created a holding company named Alphabet, which, as of this writing, has brought over 230 subsidiaries under its umbrella.[22]

The Rockefeller concept of using a trust to thwart competition soon became the capitalist rage. The Tobacco Trust, Beef Trust, Copper

Trust—even the Writing Paper Trust—all followed the acquire-and-shut-out-competition model. Likewise, Google is not alone today in its use of acquisitions to control competition and grow revenue. Facebook (now Meta) locked out competitive threats by acquiring Instagram and WhatsApp along with ninety other companies.[23] Amazon has increased its dominant market position by acquiring eighty-nine companies.[24] Sounding like a twentieth-century Rockefeller, Facebook founder and CEO Mark Zuckerberg explained in a 2008 email, "It is better to buy than compete."[25]

Destruction of Small Business

The original Gilded Age saw small businesses plowed under by the scope and scale economies of industrial production and the network effects of the railroad.

In 1878 when Gustavus Swift developed the refrigerated railcar he created the technology that would underpin the Beef Trust. Slaughtering at scale in Chicago abattoirs was significantly less expensive than sub-scale local butchers doing the same thing. Added to this was the savings from transporting only the edible cuts of beef rather than the whole cow. By offering the same product at a lower cost, Swift undercut the business of local butchers, driving many out of business and eliminating local jobs in the process.

Today, dominant digital companies still enjoy scope and scale economies. They are further advantaged by how the internet has super-charged network effects with usage-created digital information. Distant digital platform companies now know more about consumers in a locality than local media or merchants know about their neighbors. Swift's product was processing meat. Google's product is processing data about users to compete with local media. Amazon's product is knowing what we like and quickly delivering products that formerly required a trip to the local store.

The dichotomy of destroying a cornerstone of local economies while improving products, convenience, and price reverberates from the nineteenth century to today. The digital platforms have done to local businesses what Swift did to butchers.

Tempo of Life

In 1881 a neurologist identified a new generation of stress-related maladies that he ascribed to the pace of the new era. In *American Nervousness: Its Causes and Consequences*, George Miller Beard attributed what he called "nervous exhaustion" to the effects of "modern civilization," which he defined to include "steam power, the periodical press, the telegraph, the sciences, and the mental activity of women."[26]

While the last factor produces modern head-scratching, the other causes of angst have digital equivalents. The internet delivers goods and services with unprecedented speed and low cost. The periodical press has been replaced by the continuous online barrage of sensational news and information. The binary signals of the telegraph have evolved to become the language of computers and the internet.

A characteristic of both eras is the shrinkage of the time buffer that previously permitted the gradual acclimation to a new technology and its effects. In the original Gilded Age, the pace of industrial-driven change was greater than that of the preceding agriculture period. The exponential growth of the railroad—from 30 miles of track in 1830 to 30,000 miles in 1860—forced communities that had always been isolated to adapt. The telegraph running alongside the railroad further accelerated the flow of information and economic activity. Yet the pace of that change, while faster than before, still proceeded at a deliberate speed that allowed time to adjust and evolve.

Perhaps the ultimate example of the accelerated pace of change between the two Gilded Ages is the comparison between one of the first Gilded Age's notable innovations, the telephone, and its twenty-first-century equivalent. It took 125 years for Alexander Graham Bell's invention to reach one billion people. The new Gilded Age's equivalent, the Google-owned Android mobile phone, took less than six years to reach the same milestone.[27]

Adapting business and personal activity to the telephone had proceeded at a slow linear pace. The changes imposed by a connected computer in your pocket introduced change at an exponential pace. The result further accelerated the tempo of life and virtually eliminated the previously existent adoption/adaptation time buffer. A twenty-first century of "nervous exhaustion" was the consequence.

Fake News

"There is truth and there are lies. Lies told for power and for profit," President Biden observed in his 2020 inaugural address.[28] The first reference was to Donald Trump. The second was to the digital platform companies that dominate the flow of information.

Journalism has always been about selecting information to be conveyed. In a capitalistic economy that means selecting information that will maximize revenue. Today we are in the third iteration of the news curation model.

In the first model, early publications often curated news with an ideological slant. Newspapers saw no shame if their reporting reflected the views of the political party or movement with which they were associated. In fact, while they carried some advertisements, these businesses counted on their slanted coverage to entice like-minded individuals to pay six cents to read their product.

The mechanization of the Gilded Age helped reshape the news business to ultimately create the second model. It began with the telegraph's ability to quickly bring news from afar to expedite the flow of information and expand the definition of news beyond coverage of local events. Then, new powerful printing presses increased the physical size of the newspapers, creating more room for advertising designed to reach the nation's growing affluence.

The new technologies also meant the economic model changed. Instead of charging six cents a copy, the "penny press" was born: enticing more readers through lower price and thus driving the greater revenue potential of advertisements. While slanted news, such as the period's "yellow journalism," was still prevalent, ultimately, biased news had to yield to efforts that maximized revenue by promoting veracity and balance so as to offer more eyeballs to advertisers by offending as few as possible.

The new Gilded Age represents the third model of information curation. While the economic model is still about maximizing revenue, it is no longer about the need for balance and veracity. Like the early ideological media, the new media profits by playing to users' preferences and prejudices. The difference this time is that software algorithms organize the information to deliver what each user likes in order to hold the user's attention to see as many revenue-generating ads as possible. In the world of social media this means targeting users with what they want to see and hear. In the emerging AI-driven environment this can mean answering questions with what the user wants.

Perhaps the most pernicious difference of the digital model, however, is its secrecy. Unlike the previous models, which built around the public dissemination of information to a broad audience, the new business model delivers content in secret so no one knows specifically who is receiving what. The digital platforms claim they are "just a technology," but that technology makes money through the most sophisticated—and secret—content curation ever devised.

The digital platforms have also developed another revenue stream: helping purveyors of untruth get paid for their lies. While profiting through the manipulation of information has existed throughout history, those profits were traditionally limited to the information purveyors themselves. Today, major online platforms such as Google make money not only by disseminating news to their users, but also by delivering advertising—and its revenue—to known fake news sites, thus keeping those sites alive. Researchers have found that nearly half of all advertisements funding "fake news" sites (i.e., sites focusing on pseudoscience or outright lies) and "low credibility" sites (i.e., hyper-partisan sites) are fed from Google ad servers. Google thus profits not only from misinformation on its own platforms, but also from delivering ad-driven revenue to help fake news sites profit from their own untruths.[29]

The Politics of Distraction

Gilded Age historian Jack Beatty poignantly described the politics of the era as a mystery in which "reverse alchemy transformed mass enthusiasm into policies disfavoring the masses."[30] Through the political exploitation of cultural, racial, and religious issues, Gilded Age politicians were able to generate the votes necessary to get elected. Once in office, however, they ignored the abuses heaped on those voters in favor of benefiting wealthy corporate interests.

Modern politics' focus on cultural, racial, and religious issues thus resembles the practices of the original Gilded Age. This time, however, the data that digital platforms collect and hold about American citizens' prejudices and preferences allow the microtargeting of those messages to divide, distract, and manipulate the electorate even more.

Digital distraction has helped deliver a dysfunctional twenty-first-century Congress. As the digital platform companies profit by the deconstruction of society into ever more granular subsets, they dismantle the community necessary for democracy to function by destroying the connection of a common story in favor of targeted and emotionally charged topics.

The result is a similar "reverse alchemy" favoring the powerful. As political scientists Thomas Mann and Norman Ornstein have observed, today's political responsiveness to business—as opposed to those affected by business—"more closely resembles the House [of Representatives] of the nineteenth century than that of the twentieth, of the Gilded Age more than the Cold War era."[31]

Similarities versus Replication

While we have just seen some of the similarities between the original Gilded Age and the present, this is not to suggest that our Gilded Age is a carbon copy replication of the specific experiences of the mid-nineteenth and early twentieth centuries. Today's developments are, for instance, de-industrializing compared to the Gilded Age's industrialization.

The greatest dissimilarity, however, is that the exploitative practices of the original Gilded Age were eventually curbed by We the People acting through a representative government while today's abuses remain unresolved and largely unaddressed.

THE FROG ON THE FENCEPOST

Both the original and follow-on Gilded Ages relied on new networks. In the nineteenth century it was the railroad and the telegraph. In the twenty-first century it is the internet. All three networks were private entities made possible by government support.

The Gilded Ages are symbolized by individuals—from Rockefeller to Swift and Zuckerberg to Bezos. The technical assets that allowed these individuals to achieve their accomplishments were underwritten by the people who were subsequently exploited by the technology's use.

Railroads benefited from state charters that transferred the government's eminent domain power to the companies to permit the forced acquisition of land for rights-of-way. Then, when it came time for construction of the transcontinental railroad, the government transferred vast public assets to the private companies. First, the proceeds of $100 million in new federally issued bonds were directed to the companies (equivalent to over $2.5 billion today). As if that were not enough the railroads then received huge grants of public lands. The line from the west, the Union Pacific, received land equal to the size of the states of New Hampshire and New Jersey combined. The line from the east, the Central Pacific, received land equivalent to the state of Maryland.[32] This land was not just for the rights-of-way, but also to be sold and otherwise monetized for the benefit of the companies.

The federal government also supported development of the telegraph. The test line that delivered "What hath God wrought" was entirely paid for by a federal grant of $30,000 (over $1 million today) to painter-entrepreneur Samuel Findley Breese Morse.[33] Seventeen years later, Congress followed with the Pacific Telegraph Act of 1860, appropriating $40,000 for a commercial transcontinental telegraph line.[34]

Just as the federal government's investment enabled the railroad and telegraph, the internet also benefitted from government policies. The key technologies that created the tech titans were the product of focused public policy actions.

The packet data concept that is the functional heart of the internet was developed by RAND Corporation researcher Paul Baran under a Department of Defense (DOD) grant.[35]

Another DOD project, from the Advanced Research Project Agency (ARPA), built the first iteration of the internet by connecting the networks serving the computers at four U.S. research institutions.[36]

The software algorithm that became Google was developed by Sergey Brin and Larry Page while working on a Stanford University project funded by the National Science Foundation.[37]

The computing power of microchips that is so essential to the digital era enjoyed multiple boosts from the government. It was antitrust enforcement against AT&T that opened up the essential transistor patents the company controlled.[38] Government contracts were then a key to the early demand for the microprocessors utilizing that transistor technology.[39]

The lingua franca of the internet, TCP/IP, was developed by Vint Cerf and Robert Kahn while working at the Defense Advanced Research Projects Agency (as ARPA had been renamed).[40]

The first operational installation of a cellular telephone network was paid for by the U.S. Department of Transportation (DOT) on the Amtrak Metroliner.[41]

Every underlying technology in the smartphone can be traced to development supported by the U.S. government and its citizens.

"It's like the frog on the fencepost," I observed as a member of the Democratic National Committee's platform drafting committee in 2012. "They [the new companies] didn't get there by accident." Just like the frog did not get on the post by himself, "so much of what we [as a nation] are able to do as innovators and entrepreneurs we are able to do because of what we collectively have done as a society . . . all of which were seized upon by creative and innovative entrepreneurs to build good and great businesses. But all of which began with the collective of all of us working together."[42]

The innovators and risk-takers who built the technologies of the Gilded Ages should rightly be recognized and rewarded. Their activities, however, were not a technological immaculate conception. It all began with an assist from We the People.

WE THE PEOPLE CAN RETAKE CONTROL

The new government-assisted technologies produced wonderous new products and services while also creating great economic and political power for the corporations that took advantage of the public's investments. In an epic struggle in the late nineteenth and early twentieth centuries, the corporate interests fought the assertion of public interest oversight of their activities. A similar epic struggle is ongoing in the new Gilded Age. Like its predecessors, this struggle to prioritize public interest over private interest will not be resolved quickly.

As cartoons from the original Gilded Age attest, the dominant corporations of the time viewed government as something else to be owned.[43] Even the business-friendly *Chicago Tribune* complained, "Behind every one of the portly and well-dressed members of the Senate can be seen the outlines of some corporation interested in getting or preventing legislation."[44]

The Bosses of the Senate, by Joseph Keppler.
PUCK, 1889, FINE ART IMAGES / HERITAGE-IMAGES / TOPFOTO

It was typical for members of the House and Senate to receive corporate largesse ranging from cash to free railroad passes. Gilded Age U.S. president Rutherford B. Hayes confided to his diary the observation "government of the corporation, by the corporation, and for the corporation."[45]

Against these corrupt blandishments ran a populist tide, rooted in small farms and businesses, that was distrustful of big business. The corporations had the cash and clout, but the people were beginning to organize. One of the earliest examples of a populist grassroots movement against corporate power was the 1867 founding of the National Grange of the Patrons of Husbandry as the voice of farmers.

The principal focus of the Grange was the abusive practices of the railroads. Because small agricultural communities rarely had more than one rail line, that company was able to charge all the market would bear—what economists call "monopoly rents." A short haul from a rural town to a rail hub such as Chicago might cost twice as much per unit of weight as would a longer haul across a route with competitive carriers. The Grange initially focused on the need for each state to have a railroad commission to protect shippers. A state-by-state patchwork of policies, however, proved difficult to implement for an interstate service such as the railroads. Of even greater consequence was the railroads' political power, which effectively neutered these commissions.

It was a twenty-year struggle before Congress created the Interstate Commerce Commission (ICC), and a national policy to oversee the railroads. The new commission made history as the first independent federal regulatory agency, but it suffered from many of the regulatory capture problems of the state railroad commissions. It took another eighteen years—and Teddy Roosevelt—before the ICC had sufficient authority to provide meaningful oversight.

The Populists had brought the regulatory issues to the forefront; the Progressive movement pushed it forward. Whereas Populists were skeptical of big enterprises, Progressives accepted the new industrial model and sought to regulate its behavior. Together, Populists and Progressives delivered a trifecta of regulatory reforms: overseeing the networks, regulating the products those networks carried, and protecting a competitive market. The regulatory issues of today break down in a similar three-part manner.

Networks

The policy debate about regulating the railroad as the essential network of the nineteenth century previewed the policy issues for regulating the essential network of the twenty-first century: the internet.

In his December 1904 address to Congress, Theodore Roosevelt took aim at the railroad companies. "Above all else," he declared, "we must strive to keep the highways of commerce open to all on equal terms."[46] The railroads should be "common carriers" required to carry traffic on a nondiscriminatory basis for just and reasonable terms and conditions.

That an essential network should be "open to all on equal terms" that were just and reasonable was exactly the net neutrality issue of the early twenty-first century. After being debated for over a decade, common carrier requirements were placed on the providers of internet connections by the Obama FCC that I chaired. The Trump FCC, at the request of the companies, subsequently removed the requirements. The issue, however, will not go away and remains a classic example of how the development of regulation has the twists and turns—and length—of an epic novel.

Products and Services

Policies regarding the products and services the networks enabled were another focus of Gilded Age reformers with relevance to today.

The effect of Gustavus Swift's refrigerated railcar, for instance, industrialized the slaughtering of animals. The raw material was brought to a central point by rail and converted to a finished product, which was then shipped out by rail to a waiting market. The absence of oversight of the slaughterhouse process, however, introduced threats to the safety of the final product. In 1906 Roosevelt signed the Pure Food and Drug Act, establishing regulation to protect the safety of the food supply.

The effect of network-based digital platforms has similarly introduced new threats to the public interest. The invasion of privacy and dissemination of misinformation and hate are but two examples. Yet to date there has been no "Pure Digital Platform Act" to address the public interest abuses of the digital platform companies.

Competition

The third leg of Gilded Age reforms dealt with corporate concentration and monopolization.

The concentration of industrial activity among a few providers soared as merger mania swept through industries. From 1895 to 1904 there were 2,274 mergers among American manufacturing enterprises. The result was 157 remaining companies that typically dominated their respective industry.[47] Congress had passed the Sherman Antitrust Act in 1890, but as these statistics show, industrial consolidations continued at a fever pace in the years after passage.[48]

In 1914 Congress addressed the broader issues of effective competition with the Clayton Antitrust Act.[49] The new law expanded powers of the Department of Justice over, among other things, price discrimination and any other action that would substantially lessen competition. The Congress also created another federal agency with competition oversight responsibility. The new Federal Trade Commission (FTC) had the authority to move against "unfair methods of competition," as well as to enforce the Clayton Antitrust Act, including reviewing mergers for their competitive effects.

The Sherman Act, Clayton Antitrust Act, and their implementation were built on industrial concepts and targeted to industrial abuses. The challenge of the internet era is that many digital practices are difficult to categorize under statutes based on industrial assumptions. For instance, while industrial activity focuses on the product per se, digital activity focuses on the data created by consumption of the product. An auto manufacturer cannot give away vehicles, for instance, but platforms such as Facebook and Google can give away their services as the catnip that allows them to collect the private information of users which can then be repurposed into revenue.

Beyond the difference between industrial and digital activities, the antitrust laws have been decreasing in power over the past fifty years. The digital revolution coincided with a revolution in American jurisprudence. Beginning in the Reagan era, courts began interpreting the antitrust statutes in terms of their impact on consumers (principally through price) rather than the impact on the competitive marketplace. As the market power of the dominant digital companies grew, court interpretations of the basic premises of the Sherman and Clayton Antitrust Acts shrank.

Issue Symmetry

There is an amazing symmetry between the technology-driven issues of the original Gilded Age and its twenty-first-century successor.

Will networks be open and nondiscriminatory, fair, and equitable?

Will the public interest be protected against harms from the products carried by those networks?

Will a competitive market be protected and promoted?

An important issue for today is whether there will also be symmetry in the length of time required to resolve these issues. Amid today's much more rapid pace of technological advancement can we afford, for instance, the two decades of abusive practices that preceded creation of the ICC and the further almost two decades before it had the necessary powers?

THE GILDING

The behavioral similarities of the mid-nineteenth, early to mid-twentieth, and early twenty-first centuries certainly fit the thought attributed to Mark Twain that history rhymes. More concretely, today's experience contains many of the attributes that prompted Twain to dub the earlier era the Gilded Age.

The common component of today and the original Gilded Age is the gilding itself: a thin layer with a glittering patina that conceals a less than golden reality. Such gilding begins with self-advantaging practices that ignore the rules that previously protected the public interest and provided stability.

The industrial barons of the Gilded Age, following basic human nature, took advantage of a market without rules to construct a code of conduct that benefited themselves. After years of struggle, the barons' abusive behavior was finally curbed by government regulation.

The first few decades of the digital era have witnessed a similar gilding. Entrepreneurs, acting often without consideration of the consequences of their actions, are enabled to make their own rules by the inability of government to deal with their creations. Once again, new technology has produced a "marvelous material well-being" in the form of innovative new products and services. Accompanying the wonders of such technology-based services, however, is a lack of meaningful oversight of the human instinct and economic impetus for excess. There are no guardrails to protect consumers and competition from the economic and political power amassed by the technology-based companies of the twenty-first century.

An address by President Roosevelt in January 1906 could be lifted word for word to describe both today's policy challenges as well as the solution. "Neither this people nor any other free people," TR warned, "will permanently tolerate the use of the vast power conferred by vast wealth, and especially by wealth in its corporate form, without lodging somewhere in the government the still higher power of seeing that this power . . . is also used for and not against the interests of the people as a whole."[50]

We are presently engaged in the search for such a "higher power." At the core of the digital Gilded Age is the same question that arose in the industrial Gilded Age: Will there be rules for the new economy, and who will make those rules, the people or the powerful?

This Is NOT the
"Fourth Industrial Revolution"

PUBLIC INTEREST OVERSIGHT OF THE DIGITAL ERA CANNOT RELY ONLY on assumptions that were applicable in the industrial era. To define tomorrow in terms of what we knew yesterday is a very human instinct, but one that will serve us ill in the new era.

The original Gilded Age was built by the industrial revolution. The new Gilded Age is the result of de-industrialization.

To conflate the causes, effects, and policy decisions of the industrial revolution with what is happening today is to gloss over the seminal differences between the technological and economic outputs of the two eras and start down the wrong path for considering remediation of the problems of the digital economy. Identification of remedies to the problems of the digital era must begin with a realistic diagnosis of the roots of those challenges. The trick in such analysis is to avoid the tendency to define the new by what we understand about the old.

Industrialization was the investment of capital in mechanization that lowered per unit costs by increasing the output of a fixed number of workers. At its nineteenth-century core were two interrelated technological developments: steam-driven mechanical production and steam-driven transportation.

Today's de-industrialization has reduced the cost of digital goods and services even further to approach zero marginal cost. This, also, is built on two interrelated technological developments: the low-cost manipulation of digital information by microprocessors, and the low-cost transportation and manipulation of that information across high-speed digital networks.

Mark Twain's historical rhyming definitely exists between the industrial revolution that defined the nineteenth and twentieth centuries and the twenty-first-century information revolution. These rhymes inform us, but they should not instruct us. While the poetry of the eras may be similar, dealing with the challenges of the present era requires recognition of their differences.

BEYOND THE "FOURTH INDUSTRIAL REVOLUTION"

As the Gilded Age flourished in the United States, British historian Arnold Toynbee popularized the term "industrial revolution" in his *Lectures on the Industrial Revolution of the 18th Century in England.*[1] The historian's discussion was less about mechanization per se than "the substitution of competition for the medieval regulations which had previously controlled the production and distribution of wealth."[2]

The competitive market Toynbee heralded was driven by the harnessing of externally powered mechanical devices to enhance the output of human laborers. In the industrial era's earliest iteration, machines were powered by the flow of water, thus constraining their geographic location. The harnessing of steam—a power known by the ancients, but insufficiently powerful for mechanical application until Thomas Newcomen and James Watt—removed those siting limitations making production possible virtually everywhere. Then Richard Trevithick's harnessing of "strong steam" put steam on rails to conquer geography altogether. It was the productive power of steam-powered machines and the geography-conquering iron horse that created the transformational industrial revolution.

The introduction of electricity as a source of power, and the productivity improvements of the assembly line have been described as the "second industrial revolution." When the use of mainframe computers—and ultimately personal computers—became widespread in the mid-twentieth century, it was proclaimed the "third industrial revolution."

Now comes the World Economic Forum (WEF) and its founder and executive chairman, Klaus Schwab, to popularize the idea we are living through a fourth generation of the industrial revolution.

The ideas postulated in Schwab's book, *The Fourth Industrial Revolution*, are a thoughtful and challenging look at the impact of technology in the twenty-first century.[3] The book's observation, "Technology is not an exogenous force over which we have no control,"[4] is at the heart of the ideas in the book now in your hands. The concern "that decision makers are too often caught in traditional, linear (and nondisruptive) thinking"[5] is a spot-on characterization of forces that have hindered the development of meaningful modern public interest tech policy.

The trap in the assumption of a fourth industrial revolution, however, is the supposition that today is an *industrial* revolution, or the extension of industrial concepts by other means. It is particularly risky to attempt to address today's new challenges, let alone plan for tomorrow, based on such a backward-compatible assumption.

We are not simply living through a variation of the industrial revolution. The information revolution is something that is more on par with the transformation represented by the move from the agrarian era to the industrial era. The agrarian revolution, built on the domestication of crops and animals, ended humankind's hunter-gatherer existence. The industrial revolution then introduced mechanical devices that reduced production costs, pulled people off the farm into factory cities, and made widely available previously unavailable products. The industrial revolution defined human experience through most of the twentieth century. The information revolution defines today and tomorrow.

The information revolution has not extended the industrial era's centralized, factory-based activity but replaced it with a distributed, less hierarchical economy. One must only look at how this transformation played out in the COVID-19 pandemic to see its transfigurations. Work from home replaced the centralized workplace that sprung from industrialization. Industrialized education in factory-like schools became virtual and the classroom anywhere. Health care no longer required a trip to the doctor. While the new normal may end up being a hybrid of place-based and virtual activity, it is never going back to the centralized command-and-control of the industrial era.

There are five factors that define why we are not in a fourth industrial revolution. (1) Digital assets are different and behave differently than industrial assets. (2) Industrial output grew linearly, digital output is exponential. (3) Production characteristics are different. (4) The networks that create the digital connections have themselves changed. (5) And finally, because of these differences, the management of digital activities is different from that of industrial activity. Together, these digital economy realities underpin the need for a new approach to the oversight of twenty-first-century economic activity.

Different Assets Behaving Differently
The policy and practices of the industrial era were based on business activity that utilized hard assets such as coal, iron, and manufactured goods. The digital era is built around soft assets expressed in computer code. While such data assets enjoy industrial-like scope and scale economies, they are different in many other ways. Compared to industrial assets, information assets are incrementally inexpensive, inexhaustible, iterative, and non-rivalrous.

Incrementally Inexpensive—Computers manipulate data and networks distribute the results at very low incremental costs. While the computers, networks, and algorithms are (like traditional industrial facilities) expensive, their incremental output is virtually costless.

When Ford wants to build one more truck it must acquire the necessary components and put them down the assembly line at essentially the same marginal cost as the truck that preceded it. When Google does a new search, the incremental cost is a few cycles of computer time and the reuse of software files. The industrial era was driven by scope and scale economies that reduced costs. The information era is driven by marginal costs that approach zero.

Inexhaustible—In the industrial economy, a ton of coal, iron, or corn was exhausted by usage. In the information economy a file of data is inexhaustible and can be used repeatedly.

The steel that built the chassis of the incremental truck was totally consumed by that usage. When Microsoft downloads a new purchase of Word, however, it simply copies a file of digital information, leaving the original file available for subsequent use again and again.

Iterative—The use of industrial assets created a product that was typically an end in and of itself. The Ford truck, for instance, does not produce new trucks. In contrast, the use of digital assets creates new digital assets. Digital products, themselves the assembly of various pieces of data, create their own data through usage. That new data can be used in turn to create a new product.

Going online to a social media site such as Facebook, for instance, generates a new asset: the digital record of the user's activity. Facebook then iteratively uses the newly created information for a new product: the improved targeting of consumers for advertisers.

Non-rivalrous—The steel used by Ford to build a truck is not available to GM to build a truck of its own. Economists describe this as a "rivalrous" asset in that if it is used by one party it cannot be used by anyone else.

Digital information is "non-rivalrous" because it can be used by multiple parties again and again without decreasing its usefulness. A video file used by YouTube could (if corporate policy would allow) be played on Facebook with no decrease in its future functionality to either party.

New Factor of Production—An economy based on the new asset of digital information is driving the reconceptualization of production itself. Traditionally, the inputs needed for the creation of a good or service have been defined as land, labor, capital, and entrepreneurial ability.[6] It was this industrial-based definition that my generation learned in business school—but it is "so twentieth century."

In the development of its twenty-first-century master plan, the Chinese government recognized the uniqueness and importance of digital information by formally recognizing a "fifth factor of production." In April 2020 the State Council, the government's top administrative body, formally declared data to be an essential factor of production, alongside land, labor, capital, and technology.[7]

EXPONENTIAL, NOT LINEAR

The industrial revolution won the Civil War, World War II, and the Cold War by harnessing the factors of production to outproduce the opponent. The scope- and scale-based economics behind these events were a linear force. The information revolution is exponential.

The efficiencies of industrial production eventually hit a ceiling. At some point, scope- and scale-based efforts begin to experience dis-efficiencies that cause the benefits to slow. The larger an enterprise grew, the more complex and difficult it became to manage, and the resulting bureaucracy was less entrepreneurial and more sclerotic. As a result, the linear growth in efficiency and output began to tail off.[8]

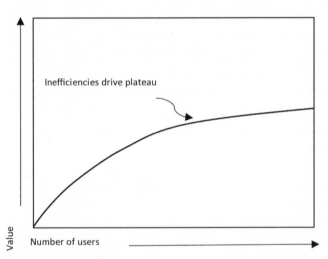

Traditional Economies of Scale

MARCO IANSITI AND KARIM L. LAKHANI, "COMPETING IN THE AGE OF AI," *HARVARD BUSINESS REVIEW*, FEBRUARY 2020.

Industrialization was about reducing production costs in order to drive profits. When those efficiencies began to wane, it produced two results—one good and one bad.

The positive result was that the plateauing of productivity created space for competitor companies. If, for instance, efficiencies began to decrease at 30 percent market share, opportunity was created for three or four competitors whose efficiencies had not peaked. Thus, while the industrial economy may have crushed small companies with weaker-scale economies, it also had a built-in governor that did not totally destroy markets.

The negative aspect of the plateauing of efficiencies was that it encouraged corporate executives to switch priorities from increasing profits by lowering costs to increasing profits by lowering competition. If there was little or no competition, prices could be set at what the market would bear. Thus the Gilded Age that began with production efficiencies ended in a flurry of anti-competitive combinations.

Data-driven economies avoid industrial plateauing. In a digital company, scaling is not so much the addition of people and machinery as it is simply connecting across a low-cost network with digital activities enjoying low marginal costs. In an era when algorithms recommend books at Amazon, select drivers at Uber, and match information with users on Facebook the processes that defined industrialization are disintermediated by build-it-once, use-it-repeatedly software code. After foundational investments in computers, connections, and code, digital economics leap past linear improvements to grow exponentially in their delivery of value.[9]

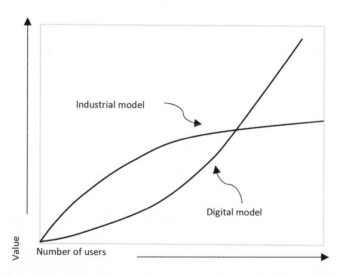

Exponential Digital Scaling
MARCO IANSITI AND KARIM L. LAKHANI, "COMPETING IN THE AGE OF AI," *HARVARD BUSINESS REVIEW*, FEBRUARY 2020.

The economics of digital goods are further enhanced by the network effects of their interconnection. The term "network effects" describes how the value of a network increases as the number of users increases. It is a kind of digital perpetual motion machine. Users go to Facebook, for instance, because that is where everyone else is. They go to YouTube because that is where others put their videos. Each new user in turn creates digital gold—their personal data—that the platforms mine to drive the targeting that is sold to those seeking to reach individual users with precision. Users attract users . . . which attracts users . . . which generates data . . . which creates a new corporate asset: a psychographic portrayal of each user that can be made available for a price to companies desiring to reach a target audience.

Because digital network effects drive usage and the creation of new digital assets, they tend to produce winner-take-all outcomes. The industrial barons of the original Gilded Age eliminated competition through corporate consolidation using trusts and other vehicles. New Gilded Age internet barons accomplish their market dominance by combining exponential digital economics with network effects to control both the collection of the raw data asset as well as its monetization.

Pairing Platforms, Not Pipelines

Google and Apple each maintain one foot in the practices of the industrial era and the other in the internet era. The production of an iPhone- or Android-based device is an industrial assembly process. The operation of the Apple App Store or the Google Play store, in contrast, is a digital platform. A pipeline produces, a platform pairs.

Industrial production is a linear procedure. Producing the smart device is a matter of moving it down a production line where functionality is added in a step-by-step process until the completed product rolls off the line. The assembly of the components is also a pipeline, a supply chain of the necessary inputs each themselves the result of a production and delivery pipeline. It is essentially the same process by which Ford produces trucks or Kellogg's produces corn flakes. It is the process that built the industrial era and the Gilded Age.

Smart devices, however, are definitely non-industrial in how they function. Rather than a pipeline, the creation and delivery of digital products results from the pairing of data assets. In place of pipeline production, the Google and Apple stores pair app developers with app users (as does the pairing of Apple Music and YouTube videos).[10] Facebook pairs users' profiles with advertisers' targets. Google pairs queries with answers. Amazon pairs people and products. As we will shortly see, artificial intelligence is also a pairing activity.

The consumer-facing platform companies operate what economists call a "two-sided market," meaning that the platform plays on both sides of the market equation. On one side, it acquires personal data, with free services as the bait. On the other side, it monetizes the use of that data by pairing consumers with each other, news, or products.

The acquisition side of the equation is aided by both direct and indirect network effects to produce the Holy Grail of the digital economy: highly granular data enabling highly precise conclusions. Facebook enjoys the direct network effects of users wanting to be on the same service as their friends. Google enjoys indirect network effects; while other search engines can answer typical queries, they do not have access to the vast amount of data that Google does for more obtuse queries.[11]

It is on the monetization side of the equation that the precision granularity of the network effects really pays off. The reason why platforms tend to be winner-take-all is because feeding the vast amounts of data produced by network effects into algorithms produces precise predictions that companies without access to such a huge database cannot match. As the only party possessing such precision, the platform can pretty much charge what the market will bear.

In the industrial pipeline economy, high margins tended to attract competition since the raw materials necessary to begin production were readily available. In the platform economy, however, acquisition of the raw material is thwarted by network effects as well as how the platforms hoard the data. As a result, advertisers, and other users of the platform's targeting, are confronted with few choices if they want the benefits of top-quality data.

NETWORKS: FROM TRANSPORTATION TO ORCHESTRATION

Even the railroads of the digital era—the digital networks—have become platforms. The pathway of the railroad—or telegraph, or analog telephone—was a "dumb" point-to-point circuit to be filled with someone else's content. The pathway of a digital network is a "smart" series of computers that exchange information as they pass data.

As microchips are built into everything, wireless connections allow access to the information the chips create as well as the functions they control. Earlier iterations of the internet created a product through a call-response that requested and then displayed information. The new web orchestrates intelligence to create a product.

The evolution of networks from the *transportation* of a fixed amount of data to the *orchestration* of a flood of intelligence from connected microchips is the final nail in the coffin of the "fourth industrial revolution."

Consider, for instance, the difference between a "connected car" and an "autonomous vehicle." The connected car is the digital version of an industrial network that delivers a product from one point to another. The car may be online and able to read emails to you, but it is still the basic point-to-point of the old-style network: the request-reply of information out of and into the automobile.

Autonomous vehicles, in contrast, are chock full of microchips generating intelligence that must be orchestrated with data coming from other vehicles, road signs, weather sensors, and a myriad of other connected microchip inputs. That orchestration occurs both in the network and in the vehicle. The production of the new product—the safe coexistence and cooperation of vehicles—is not an industrial activity, but an information activity. It is the redefinition of value creation from a network acting as a pipeline to a network acting like a platform to pull the intelligence from ubiquitous microchips to be manipulated for the creation of new products and capabilities.[12]

ANTI-HIERARCHY MANAGEMENT

As the industrial revolution produced large operating entities, the management techniques that had previously sufficed for small enterprises were no longer appropriate. When two blacksmiths working their way through eleven separately identifiable tasks could produce a plow in approximately 118 man-hours, management was simply a matter of communication between the two. On the factory floor, however, that same product was the result of harnessing 52 men, performing 97 distinct tasks, to produce the plow in just 3.75 man-hours. The involvement of so many individuals across so many tasks required a hierarchy of management supervision to oversee and track the activity.[13]

Industrial management was a legacy of the railroads. The construction and oversight of railroads was the largest project in human history. Not surprisingly, the railroad companies turned for management expertise to the nation's only other large-scale entity, the U.S. Army. Approximately 120 graduates of West Point became senior railroad executives, while the subordinate ranks were filled with countless others with previous army experience.[14] The result was centralized, command-and-control management. (Ever wonder how military terms such as "division" came to be applied to corporations?)

In 1846, the B&O Railroad issued its *Proposed New System of Management*, which established functional chains of command.[15] The new model was soon embraced by other railroads and then by non-railroad industrial enterprises. As scope and scale industrial production harnessed masses of workers to deliver to a national market, corporate bureaucracy provided supervision and structure. Introduction of the telegraph then allowed supervision of multiple locations from afar, resulting in yet another layer of management.

Digital networks stand such management hierarchy on its head. The production of hard capital assets in centralized locations enabled a stratified corporate structure; information-based activity works in the opposite direction. As the location of work moves off the shop floor into the distributed network, coordination via hierarchy is replaced by coordination via clicks to links and apps.[16] Of course, there is still oversight, but that oversight is based more on results than on rules.

As COVID-19 proved, it is not necessary to assemble at a central place to perform many corporate functions. And because digital platform products are pairing activities as opposed to pipeline activities, the inputs to such pairing can be widely distributed. Consumer-facing platforms such as Facebook and Amazon pair people and products "in the cloud"—something that is not a place per se, but the graphic representation of the interaction among distributed computers. Company-based platforms perform a similar pairing among employees and clients to produce the end product.

Such pairing activities are by definition anti-hierarchical. In that regard, they begin to resemble the interaction of the two blacksmiths of the pre-industrial era: a self-enforcing, self-delivering activity. Such anti-hierarchical activity brings management full circle back to the artisanal economy that was replaced by the industrial economy.

CHAPTER 3

Closing the Open Internet

YOUNG (29) ALEXANDER GRAHAM BELL'S TELEPHONE WAS ONE OF THE wonders of the Gilded Age. That did not, however, prevent it from being a technology in search of a market. Early ideas included using a phone call to alert a customer that a message had been received at the telegraph office.

The genius who built a business based on Bell's discovery was Theodore Vail. Two years after Bell filed his 1876 patent, Vail joined the Bell Telephone Company. He left in an 1889 dispute over his vision for the company. Returning in 1907, Vail built American Telephone and Telegraph (AT&T) into what would become the largest company in the world.[1]

The foundation of Vail's success was the inefficiency of telephone technology which, he argued, made it a "natural monopoly." Ever since "Mr. Watson—Come here," a telephone link required a dedicated circuit between each end of the connection. It was a highly inefficient process in which circuit capacity had to be built and then lie fallow in anticipation of peak demand. Vail argued such inefficiency meant only one company could be economically viable.

The brilliance of the internet's digital technology is its ability to overcome that inefficiency. The result also replaced the telephone's centralized and closed communications pathway with a more distributed, open, and accessible structure.

On top of the internet's open network, however, the platform companies have constructed a closed superstructure that they have exploited to

33

compromise personal privacy, destroy market competition, and disseminate lies and misinformation.

CREATING A DOMINANT DIGITAL SUPERSTRUCTURE

The digital network that underpins the internet and its applications was conceived in 1964 by Paul Baran, a researcher at the RAND Corporation. Historically, networks were a centralizing force. From animal pathways, to railways, to telephone connections, networks were organized around a central point where traffic was rerouted onward to its destination. Baran's concept—developed under a Defense Department contract seeking a communications topology that could survive enemy attack—was to format information into digital "packets" that would avoid the vulnerability of centralized switching as each packet found its way across a fishnet-like network to be reassembled at the destination.

In his paper, "On Distributed Communications," Baran used the illustration below to explain his concept.[2]

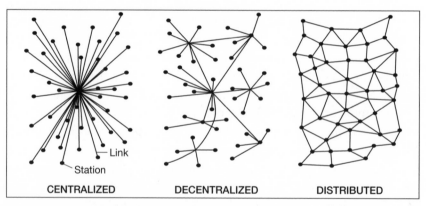

From Centralized to Distributed Networks

PAUL BARAN, "ON DISTRIBUTED COMMUNICATIONS," RAND CORPORATION, MEMORANDUM RM-3420-PR (AUGUST 1964), HTTPS://WWW.RAND.ORG/CONTENT/DAM/RAND/PUBS/RESEARCH _MEMORANDA/2006/RM3420.PDF.

The effect of distributing network activity outward to multiple switching points was to also disperse the power of the centralized network monopolies by eliminating the chokehold of the monopoly's centralized switching. When Baran's idea was presented to AT&T, the company turned it down cold. The new structure would open up their long-closed network and diffuse its power. "We're not going into competition with ourselves," was the monopoly's response.[3]

The distributed packet network changed the economics of networks. Instead of having to maintain a dedicated circuit for each connection, the packets of data were jammed cheek by jowl, thus increasing network capacity, and lowering costs.

In addition to being open by design of its topology, the new digital network is more open because of its common language and the interconnection it allows. The name "internet" is a contraction of the original "internetworking" name. What previously had been a disparate collection of networks, each using a separate communications protocol, was knitted together into a common structure utilizing a digital lingua franca called Internet Protocol (IP).

Before IP, networks and devices were "walled gardens" that not only required special coding, but also allowed the network to demand payment to access their subscribers. For instance, in the early 2000s, before wireless networks became IP networks, I was a partner in a venture capital firm that invested in a new mobile phone app. The absence of a lingua franca meant it was necessary to build the app specifically for the unique software of each device. It also meant that the mobile networks were able to control access to the walled garden in which they held their consumers and to demand a large percentage of the app's revenue before they would allow access to those consumers.

The common language of IP and the open networks of the internet tore down those walls. Also eliminated by IP was the cost of having to develop a unique coding for each device. When the networks' role as gatekeepers was eliminated, the result was the exponential growth of new and creative applications that today we take for granted.

When digital network operators tried to use their position as the consumer's link to the internet to reinstitute a chokehold on the new network, the government stepped in. The new digital networks—now called "internet service providers" (ISPs)—would, the Federal Communications Commission (FCC) ruled, be "common carriers" required to provide nondiscriminatory access to their connections. The policy was called net neutrality.[4]

The digital platform companies, having built their businesses on the new open networks, loudly supported net neutrality regulation. However, once assured the networks would not shut off the ability to reach consumers, the platform innovators started building their own walled gardens on top of the open network. The superstructure they imposed was not "the internet" but rather the undoing of the openness of the internet and the stimulation of competition and innovation it permits.

The platform companies have simply replayed the strategy developed by Theodore Vail over a century and a quarter earlier: market dominance through controlling a closed system.

THE PLATFORM CHAIN REACTION

The closed platform ecosystem begins with the harvesting of personal information. This data is then hoarded in order to maintain market control. Finally, that market dominance affects the choices available for news and information. It is a digital chain reaction with each stage reinforcing the next. The result is profits for the companies and problems for society.

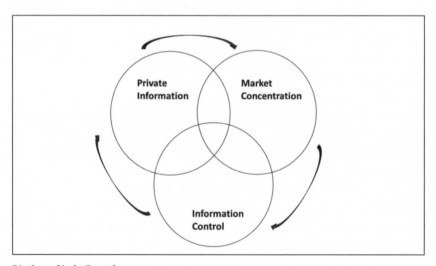

Platform Chain Reaction

Converting Personal Information into a Corporate Asset

The chain reaction begins with the invasion of the privacy of an individual's personal information. Ubiquitous microprocessors not only perform computations, but also collect information about consumers as they use the internet. Through a kind of digital alchemy, the platforms have been able to convert such private information into a corporate asset.

In authoritarian nations such as China and Russia, the government collects information about its citizens for the purpose of political power. In digital capitalism, the platforms collect information about consumers for the purpose of market power.

As will be further discussed later, control of the data asset of personal information is the platform companies' gateway to control of the internet ecosystem.

Market Concentration

Once in control of the digital asset, the companies build a moat around the data to keep it out of the hands of potential competitors.

Since data are the necessary raw material for digital services, keeping it from others allows the companies to concentrate their market control. As Federal Trade Commission (FTC) chair Lina Khan told Congress, "Control over data has enabled dominant firms to capture markets and erect entry barriers."[5]

Just as the platform companies perverted the open interconnectedness of the internet to superimpose a closed service, so they have destroyed the non-rivalrous nature of data. As we saw in the last chapter, one of the great advantages of digital assets over industrial assets is that they can be used and reused by anyone. In theory, such open availability of digital raw material should stimulate competition and innovation. In reality, the platform companies thwart competition by blocking the availability of the asset that is essential for others to compete.

Information and Disinformation

Shutting down the internet's openness and monopolizing the data that describe each user means that platform companies can determine what information their users see. Whether it is placement in a Google search or a product being sold by Amazon, the platforms' dominant market position, coupled with precise knowledge of the customer, permits picking and choosing what each user sees.

Controlling the user experience by determining what they see is particularly concerning when it comes to social media. According to a Pew Research study, 48 percent of American adults say they get news from social media sites "often" or "sometimes."[6] It is one thing for a dominant platform to decide which shoes are presented to the user; it is something far more concerning when the platform determines what news and information the user sees.

By exploiting their market and data dominance, platforms are able to control what users' see and to discourage those users from seeking information from alternative sources. As will be further explored in chapter 11, such information is essential, not only for the functioning of a competitive marketplace, but also the maintenance of a functioning democracy.

Rinse and Repeat

The data created by each user's interaction with the platforms as well as the platforms' tracking of other online behavior then feeds back into the platforms' databases to further enrich what is known about each individual and drive the cycle anew. Such refinement increases the granularity of the targeting, and the entire highly profitable process begins again as the further targeted data expands both market control and information control.

NUCLEAR EXPLOSION

Endemic to social media is how the nuclear center of the platform chain reaction can be explosive. When the three forces of the chain reaction fuse to control the open flow of information essential for a democracy to function, they can produce their most devastating results.

For social media platforms in particular, the fusion of targeted personal information and market dominance permits them to prioritize profits over probity. Of course, traditional media are also governed by the

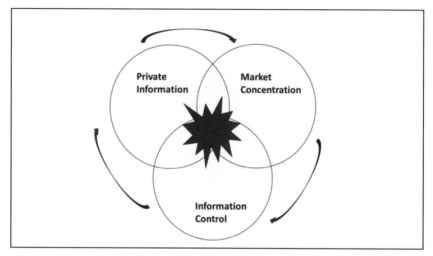

Explosive Consequences

profit motive—but implementing traditional media's profit motive has a far different result than the quest for returns from social media.

Traditionally, media companies' profits were enhanced through balanced coverage that, by offending as few as possible, attracts the most readers/viewers for advertisers. Platform companies have the opposite incentive. In lieu of providing all sides of a story to attract users, the platforms use their knowledge of each consumer to target information that appeals to that user and then sell that precision to advertisers.

The targeting of the preferences and prejudices of users extends not just to advertising, but also to the news itself. Utilizing their precise knowledge of each user, the platforms manage the message of their content to hold each user's attention through agreement or outrage. Extending the user's attention through such manipulation, allows the platform to display (and get paid for) more advertisements.

While these techniques may be profitable strategies, their practical effect is to keep away from users news and information that does not comport with the user's prevailing beliefs. As such, it strikes at the heart of the democratic process and its foundation in common information available to all.

In a 2022 interview, Facebook founder Mark Zuckerberg observed, "In order to have a cohesive society, you need to have a shared foundation of values and some understanding of the world and the problems we face together."[7] Instead of bringing us together, however, social media companies, whether Facebook, YouTube, TikTok, or others, program their algorithms to prioritize revenue by placing users into targeted cohorts that become echo chambers that are the antithesis of "shared foundational values."

The actions of Facebook during and after the 2020 presidential election demonstrated the conflict that can arise between profits and probity.

Unbeknownst to its users, Facebook had been monitoring the quality of the information it was distributing. Through a formal process, the company assigned a "news ecosystem quality" score, or NEQ, to the news sources distributed by the platform.[8] As conspiracy theories spread, management made the decision to tweak the NEQs to emphasize quality journalism. According to the *New York Times*, "The change resulted in an increase in Facebook traffic for mainstream news publishers including CNN, NPR, and The New York Times, while partisan sites like Breitbart and Occupy Democrats saw their numbers fall."[9]

These efforts to defuse the explosive power of the platform did not last long. By the end of November, the old algorithm that distributed lower-quality, more inflammatory content was back in place. As Cecilia Kang and Sheera Frenkel reported in *An Ugly Truth: Inside Facebook's Battle for Domination*, there were fears "that Facebook's changes had led to a decrease in sessions: users were spending less time on the platform," with a resulting effect on advertising revenue.[10] "The bottom line was that we couldn't hurt our bottom line," one Facebook data scientist told Kang and Frenkel.[11]

Following the January 6 insurrection at the U.S. Capitol Facebook conducted an after-action analysis that concluded, according to the *Washington Post*, that "in the weeks after the election, Facebook did not act forcibly enough against the Stop the Steal movement . . . even as its presence exploded across the platform."[12] Documents filed with the

Securities and Exchange Commission (SEC) by Facebook whistleblower Frances Haugen suggest that "the company's internal research over several years had identified ways to diminish the spread of political polarization, conspiracy theories and incitements to violence but that in many instances, executives had declined to implement those steps."[13]

After the experiences of January 6, the Facebook Oversight Board reviewed the record of management's behavior. Among the Oversight Board's recommendations was that "Facebook should review its potential role in the election fraud narrative that sparked violence in the United States on January 6, 2021 and report its findings."[14] The company rejected the recommendation, responding it was only "implementing in part" the findings. "The responsibility for January 6, 2021 lies with the insurgents and those who encouraged them," the company asserted in response.[15]

For a long time, Facebook (now Meta Platforms) CEO Mark Zuckerberg has argued his platform is not a media company. "I consider us to be a technology company because the primary thing we do is have engineers who write code and build products and services for other people," he told Congress in 2018.[16]

The world that digital platforms create, however, is not quite as simple. The digital chain reaction created by the "engineers who write code" has provided social media companies with a reach that dwarfs traditional national media outlets.

Delivery of news and information at scale is a media business.

Possessing huge quantities of personal information about each user turns the companies into targeted media companies.

Creating information silos does not encourage "shared foundational values" but rather echoes techniques of authoritarian state media.

When almost half of all Americans look to social media as a major source of news and information these companies are not simply "a technology company."

The dominant platform companies have exploited the chain reaction to become powerful media companies without sufficiently or appropriately exercising the responsibilities that accompany such a powerful position.

Having restructured the flow of news and information in the temporal world, these same companies are now hard at work creating a new ethereal world—the metaverse—that can restart the digital chain reaction all over again.

PART II

YOU AIN'T SEEN NOTHIN' YET!

The problems created by today's digital platforms were, for the most part, unanticipated. We cannot claim such innocence about the metaverse and artificial intelligence.

CHAPTER 4

The Metaverse

WHEN I VISITED MARK ZUCKERBERG IN 2015, HE INSISTED THAT after we were done, I must go across the street to meet the Oculus team. The prior year Facebook had paid $2 billion to acquire Oculus, a virtual reality (VR) company.[1] As I strapped on a VR headset that enabled me to fight dragons and leap tall buildings, I wondered just why a social media company would spend so much money on something that seemed so different from the online pairing of people with other people and products.

The answer came in July 2021 when Zuckerberg said on his quarterly earnings call with investors, "I expect people will transition from seeing us primarily as a social media company to seeing us as a metaverse company."[2] Three months later Facebook changed its corporate name to Meta Platforms, Inc.[3]

WELCOME TO THE METAVERSE

The future of online platforms is a pastiche of the digital sources and games we know today, enhanced by augmented and virtual reality, artificial intelligence (AI), and a chilling expansion of the invasion of personal privacy. Called the metaverse (a term coined in Neal Stephenson's 1992 science fiction novel *Snow Crash*[4] and combining "meta," meaning "transcending," and "universe"), the new platforms create a video game–like pseudo-world.

This is not a game, however, but personally identifiable avatars interacting with each other in an artificially created environment. Thus far, internet platforms have been an observational experience that principally harnessed text and video. The metaverse is an experience of participatory immersion in which real-world people, problems, and patterns come to life in an unreal world.

The metaverse brings the promise of new tools for education, entertainment, medicine, and commerce. Boeing is already talking about designing a new-generation aircraft in the metaverse.[5] Medical students will be able to go inside the human body and engineers build trial constructions. Broadway will be an immersive experience. History lectures will come alive with virtualized time travel.

At the same time as the metaverse delivers such breakthroughs, however, it also imports the problems associated with the current digital platforms while creating a host of additional new issues. The online challenges we have thus far failed to tame, such as privacy, competition, and misinformation, will be supercharged by the intrusive, immersive, individually identifiable, and manipulative nature of the metaverse. On top of this, the metaverse expands the problems we already know to be inherent in unsupervised online communities such as harassment, manipulation, threats to personal safety, and threats to the safety of children.

We are told the metaverse is years away, giving us time to figure out how to deal with its ill effects while still encouraging development of its positive features. We should not be lulled into inaction, however. Failure to develop meaningful policies now will mean we did not learn the lesson of the last multiple decades of internet exploitation in which a handful of companies protected their own self-interest at the expense of the public interest. The experience with social media is a warning about what happens when public interest expectations are not part of digital innovation.

A Meta Platforms advertising campaign promoting a softer, gentler metaverse promises "The metaverse may be virtual, but its impact will be real." That reality is presently barreling down on us without sufficient forethought about how to protect the public interest.

FROM SOCIAL MEDIA TO SOCIAL VIRTUAL REALITY
The move from the internet platforms we know today to the metaverse is a transformational move from observation to participation. Today's online

activity began as an observational experience that gradually expanded through social media and online games to become increasingly participatory. The metaverse accelerates that expansion, utilizing VR, augmented reality (AR), AI, and constant connectivity to create an immersive 3-D first person experience that puts the user "inside" a new pseudo-world.

In one form the metaverse is a move from social media to social virtual reality. In another form it is a postindustrial and a post-web means of production.

In the social media iteration, digital pioneers charged ahead to exploit the capabilities of technology without considering the consequences—and then acted surprised about the adverse effects they had created. When asked by a reporter what his company was doing "to ensure the problems of today's internet don't carry over into—or, worse, get amplified by—the metaverse," Mark Zuckerberg responded, "We have some time to try and work some of this stuff out up front."[6]

No, we don't have time. As we charge into the metaverse we have not resolved the problems of the pre-metaverse experience, let alone begun to deal with the new issues the metaverse creates.

Gartner Research forecasts that by 2026 one quarter of the U.S. population will spend at least one hour per day in the metaverse.[7] Even if the Gartner analysis is optimistic (as some reports seem to indicate),[8] the clock is ticking. The companies that hope to profit from the metaverse are moving ahead with dispatch.

Meta Platforms announced its 2023 metaverse expenditures will be $19.2 billion, or 20 percent of its total expenses.[9] According to Mathew Ball, author of *The Metaverse: And How It Will Change Everything*, the total spend in metaverse development is $200 billion annually.[10]

The ultimate effects of the metaverse will be determined by who uses the next few years better: those with private interests, or those watching out for the public interest.

TODAY'S ONLINE PROBLEMS MAGNIFIED

The issues that plagued the pre-metaverse internet—violation of privacy, domination of markets, and misinformation—do not get better because of the metaverse. Without behavioral rules for the dominant digital platforms, the new unreal world only means new company-developed rules that make the situation worse.

The Ultimate Privacy Invasion

The metaverse is a world of expanded surveillance. To be realistic, the avatars that represent us in the metaverse must know all about us. Putting on a metaverse headset will have more far-reaching and revealing results than hooking up to a lie detector. That the necessary personal surveillance is being developed by the same companies that have already trampled personal privacy should be of concern.

When the world's leading neurologists assembled in Seattle in June 2022 for that year's ACM Symposium on Eye Tracking Research and Applications, sponsors included Google and Reality Labs, a division of Meta Platforms.[11] The companies that have set out to build the metaverse have taken an interest in neurology because of the insights it provides into human behavior.

Poets say the eyes are the window into a person's soul. Neurologists are less romantic. Neurological studies have found that eye movements can reveal our thought processes.[12]

Thus far in internet history, the digital platform companies have collected previously private information about each of us by tracking keystrokes, mouse clicks, or location coordinates. All that changes when consumers don virtual reality headsets to access the metaverse and thus open the door to the collection and manipulation of even more powerful biometric information.

Today, the metaverse requires a bulky headset over your eyes. Tomorrow it will be lightweight glasses and eventually contact lenses. These devices are designed not only to harness VR and AR to deliver users to the metaverse, but also to collect biometric information about the user to feed into AI algorithms governing the avatar's actions. Eye movements, heart rate, facial expressions, even perspiration will create new data points to be measured, then manipulated in the metaverse—and monetized in the real world.

The companies' interest in neurological research is, therefore, quite understandable. After all, it was these same companies that relied on psychological studies of casino gambling behavior to develop similar techniques to make the online experience as addictive as possible. Playing off human psychology to get users to stay online for as long as possible allowed the platform companies to both sell more advertisements and collect more information on each user.

Harnessing neurological patterns can be even more powerful. Research shows that understanding how to interpret eye movements, for instance, creates data that can be used to manipulate behaviors. One study even demonstrated how eye-tracking can be exploited to influence the moral decisions people make.[13]

Metaverse devices able to collect personal information from our body's behavior are a potent new tool for digital companies. Meta Platforms has already patented technology to build eye tracking and facial movement tracking into the equipment used to access the metaverse.[14]

If we are concerned about personal data being used to manipulate people today, consider what the metaverse will allow. Instead of inserting a targeted message in a static web page, the metaverse filter will be able to select what each person sees in the virtual world. "And no, you won't just take off your AR glasses or pop out your contacts to avoid these problems," AR pioneer Louis Rosenberg warns, "Because faster than any of us can imagine, we will become thoroughly dependent on the virtual layers of information projected all around us. It will feel no more *optional* than internet access feels optional today."[15]

Metaverse advocates promise they will be responsible with our virtual world information. "[E]veryone who's building for the metaverse should be focused on building responsibly from the beginning," Mark Zuckerberg promised in 2021.[16] By any rational interpretation, however, the wishful "should be" is a far cry from "will be."

New Dominance

The metaverse opens a new marketplace to be dominated by the digital platforms. Thus far, the platform companies have conrolled the software that drives their sources. With the metaverse, these same companies are integrating into the necessary hardware as well as the sale of goods and services made possible by the software-hardware combination. Mark

Zuckerberg told his investors that while Facebook's metaverse offering will still sell advertising, "I think digital goods and creators are just going to be huge."[17] The personal information already possessed by the platforms, coupled with that which metaverse usage will generate, creates the tools by which a company can control not only social connections and entertainment, but also the virtual goods it "manufactures" at zero marginal cost out of zeros and ones. There is little reason to believe that—like the pre-metaverse world—market power will not flow to the platform with the most information about users to feed the AI that powers what the user experiences.

The control of personal data has been the key to the riches of the social media companies. After capturing private information, the companies build a moat around that data to deny access to others. The result of such data hoarding has been to thwart competition through the unequal distribution of the essential asset necessary for competition.

The major platform companies are dominant because they control the digital information on which advertisers rely, while denying it to others. The operation of the metaverse will expand the amount of private information collected to feed data-dependent algorithms. Such increased dependence thus can become a new tool used by incumbents to stifle competition and the innovation competition brings.

If the metaverse is to live up to its potential, a thousand flowers must be allowed to bloom, not just a few potted plants protected by competition-preventing practices. "No single company can or should control the metaverse," Nick Clegg, Meta's president for Global Affairs, has written.[18] Whether the metaverse is vibrantly competitive or becomes like the current noncompetitive social media landscape, however, is a matter that is too important to be left to the companies to determine.

Content Moderation and Manipulation

In today's online world toxic hate groups exploit technology to extend their bile. To expect something different in the metaverse is whistling past the graveyard.

Today's social media have been a vehicle for misinformation, disinformation, and malinformation. Editorial decisions by digital platform algorithms have become tribal political issues. The metaverse's expansion of the user's experience from observational to participatory will increase its emotional impact and accentuate the importance of editorial input.

Tech journalist Casey Newton identified such issues in an interview with Mark Zuckerberg. "Who gets to augment reality?" he asked.[19] Newton went on to imagine "a world where we're all wearing our headsets, and we're looking at the U.S. Capitol building . . . most of us have an overlay that says, 'This is the building where Congress works' . . . [but] some other people might see an overlay that says, 'on January 6, 2021, our glorious revolution began.'" Zuckerberg's response called this "one of the central questions of our time."

Social media platforms have been described as "automated propaganda" as algorithms direct information to audiences likely to respond favorably. In the Cambridge Analytica scandal, personal information ended up targeting political messaging. That messaging, however, was a lifeless product in the form of preprepared text or video targeted to a cohort of users. The metaverse, in contrast, is designed to be lifelike and personal. Reading the biometric data of a user, a propaganda avatar could be capable of adapting its look, tone, and message to fit what the sensors are reporting about the emotions of the user. "Virtual reality environments turn out to be really ideal environments for doing emotional manipulation of all sorts," RAND Corporation adjunct senior information scientist Rand Waltzman has observed.[20]

NEW META-ISSUES

Thus far, this discussion has considered the expansion of the problems of today's social media into social virtual reality. The metaverse also creates a new set of challenges: how (and whether) to import the social norms of the real world into the virtual world.

Harassment

The pseudo-world of video games already runs rampant with "harassment, assaults, bullying and hate speech."[21] A study from the Center for Countering Digital Hate found metaverse "users, including minors, are exposed to abusive behavior every seven minutes."[22]

"My nightmare trip into the metaverse" is how one reporter described the "barrage of assault, racism and rape jokes" she experienced.[23] Within minutes of entering a VR chat room, "I saw underage kids simulating oral sex on each other. I experienced sexual harassment, racism and rape jokes . . . despite using a profile that I'd listed as being 13 years old."

Research has identified the "online disinhibition effect" that leads people to behave differently online than in the real world.[24] It is no wonder, therefore, that "70 percent of women say personal safety in the metaverse is a concern, while 73 percent are worried about online abuse and 64 percent are concerned about sexual harassment."[25]

Property Rights, Financial Transactions, Fraud, Death, and Taxes
Does the combination of zeros and ones represent tangible "property"? When a digital Gucci bag was sold in a digital world for $4,100, what was being purchased?[26] Presumably, the bag was to be worn by an avatar and thus was a product, just like a real Gucci bag. While there are rules in metaverse platforms against theft, what if the virtual bag is stolen by another avatar that has hacked the target?

In 2009 a court in The Netherlands convicted three minors of stealing virtual furniture from an online multiplayer computer game by stealing the owner's identity.[27] The result has been a legal debate as to whether a non-real item can be "stolen," and if so, what criminal statutes apply. An even more basic question in the metaverse is the location of any remedy, both in the identification, apprehension, and adjudication of the offenders.

And we haven't even begun to get into Franklin's "two certainties" of death and taxes. Does homicide exist if you "kill" a personally identifiable avatar? Certainly, killing is rampant in video games; does that mean that killing a personally identifiable avatar should be morally acceptable in the metaverse? With respect to taxes, if the metaverse is going to be the new venue for online commerce and the value of those transactions is held in the metaverse rather than transferred to hard cash, what should be the taxation policy and who has jurisdiction?

Safety

In September 2022 a British coroner concluded the 2017 suicide of fourteen-year-old Molly Russell was a result of platform algorithms that had systematically shown her graphic self-harm and suicide images and videos. "She died from an act of self-harm while suffering from depression and the negative effects of online content," Senior Coroner for North London Andrew Walker concluded.[28] "The sites normalized her condition" and created "binge periods" of material that "affected her mental health in a negative way and contributed to her death in more than a minimal way."

The metaverse will also be where new lifelike relationships are established. We have already seen text-based exploitation on the web. What happens when avatars advance that to groping and other activities?

The transferability of social media effects to the social virtual reality experience is at this point only problematic, but it certainly seems logical. Blundering ahead into the metaverse unawares and unresponsive to such problems is not a safety strategy.

Discrimination

In 2019 it was discovered that the algorithm used to predict the prognosis of hospital patients favored white patients over black patients. A similar AI program used in court sentencing predicted twice as many false positives for recidivism for black offenders than white offenders.[29]

A study of structural racism in the digital revolution reported how a Google search "matched 'Black-sounding' names with the profiles of arrest records, even when false . . . [and] Facebook apologized for an AI model that asked viewers of a British tabloid video featuring Black men if they wanted to 'keep seeing videos about Primates.'"[30]

Digital algorithms have created new opportunities for discrimination, often subtle and perhaps unintentional, but discriminatory, nonetheless. The difficulties of algorithmic bias in online services have been well documented.[31] Feeding algorithms with the expanded amount of personal information gathered by metaverse companies and using that data to drive avatar behavior opens the door to the potential of platform-induced discrimination, whether intentional or not.

PLAYING CATCH UP AGAIN

"In the past, the speed at which new technologies arrived sometimes left policy makers and regulators playing catch up," Nick Clegg told the October 2021 Connect conference. "It doesn't have to be the case this time around because we have years before the metaverse we envision is fully realized."[32]

Which brings us to the theme of this book: who makes the rules?

Absent rules protecting the privacy of personal information, that information will continue to be purloined and exploited—this time in even greater depth—all to create a world that doesn't even exist.

Absent rules to assure competition and open markets, a handful of companies that hold digital information about individuals will use it not only for current market domination, but also to dominate new pseudo-worlds.

Absent rules that impose curatorial responsibility on the disseminator of information—including the creation of unreal worlds—the profit motive will continue to be the overriding rule as to what constitutes responsible balance.

For the last few decades, it has become increasingly clear that building technology simply because it is possible to do so can be destructive absent the construction of rules for that technology's behavior. Far from climbing out of the challenge presented by the first few decades of the digital revolution, the metaverse is digging the hole deeper absent meaningful rules for the behavior of the new technology.

CHAPTER 5

Artificial Intelligence

WHILE THE METAVERSE IS BUILDING A NEW ONLINE EXPERIENCE, ARTIficial intelligence (which is also key to the metaverse) is harnessing the internet's connectivity to produce answers. Basic AI has long had a role in our lives, such as when Amazon recommends a book or Google auto-completes a search query. In November 2022, however, AI developed a conversational relationship with us as it began chatting through the program ChatGPT.

Four months later, in March 2023, OpenAI, developer of the GPT algorithm, unveiled GPT-4, a program it claimed "exhibits human-level performance." To prove this, the developers had the program take exams such as the Uniform Bar Exam, the test most lawyers must pass after three years of law school. The software scored in the top 10 percent of test-takers.[1]

The tech companies whose race to build digital platforms brought us the digital challenges of today are again racing—this time to commercialize this AI breakthrough.

Mark Zuckerberg continues to advocate that the metaverse "remains central to defining the future of social connection." Nevertheless, he told his employees in the same month he made that statement, "Our single largest investment is in advancing AI and building it into every one of our products."[2]

"AI is probably the most important thing humanity has ever worked on," Sundar Pichai, CEO of Google has prophesized. "I think it is something more profound than electricity or fire."[3]

Amid all the excitement, there is also growing concern about the dark side of AI. Tristan Harris, an early Google executive turned skeptic, began warning, "What nukes are to the physical world . . . AI is to everything else."[4] In the same vein, a 2022 survey of AI experts asked, "What probability do you put on human inability to control future advanced AI systems causing human extinction or similarly permanent and severe disempowerment of the human species?" The median response was a 10 percent chance *that AI would threaten the human species*.[5]

Sundar Pichai dissents from those conclusions. "Artificial Intelligence will save us, not destroy us," Pichai believes.[6]

Regardless of which prediction one embraces, we have seen this movie before: Digital innovators rushing headlong into the development of new capabilities, making the rules as they go, regardless of the consequences for the rest of us. Like the digital experience to date, AI has the potential to create both amazing new capabilities as well as serious problems. Repeating the earlier online experience by expanding the intelligence capabilities of those same companies (and others) without the establishment of public interest guardrails is to ignore history.

To keep the corporate race to AI from being a reckless race requires governmental oversight. To keep that regulation from stifling AI innovation and investment requires something more than the replicating of industrial era concepts.

How Did We Get Here?

Artificial intelligence is based on the supposition that computing machines could replicate the neural networks of the human brain. There are over twenty billion tiny neurons in the human brain; the power of the brain comes from tying the neurons together. In the mid-twentieth century the same idea began to be applied to program computers to tie together and parse through large amounts data to come up with a pretty good guess. It was a digital pairing process to analyze a great deal of data to pair the query with an information file. The development of large language models (LLM) then incorporated the "natural language" used by people rather than computer programming language to search large amounts of digitally stored words in search of a statistically sustainable pattern of words that could be expressed as a conclusion.

To move from AI providing a pretty good guess to a high-certainty conclusion requires the trifecta of AI: the collection of massive amounts of data, access to that data by an inexpensive digital network, and the utilization of vast amounts of computing power. Those three factors converged in the early twenty-first century.

In 2018, the AI research laboratory OpenAI introduced the Generative Pre-trained Transformer (GPT). The pre-training in the name refers to how the model analyzed a huge amount of textual data to be trained to predict the next word in a specific context. String together such one-at-a-time calculations and the model produces a fulsome answer. Computer scientists took note, but the world woke up to GPT in November 2022 when OpenAI introduced ChatGPT, in which the GPT capability was structured as a chatbot to mimic human conversation. Generically, this is known as "generative AI" because it generates an answer in response to a query.

Seemingly overnight, ChatGPT's answers to questions went viral. One of the early viral examples was to ask the software to explain how to remove a peanut butter sandwich from a VCR, but do so in the language of the King James Bible:[7]

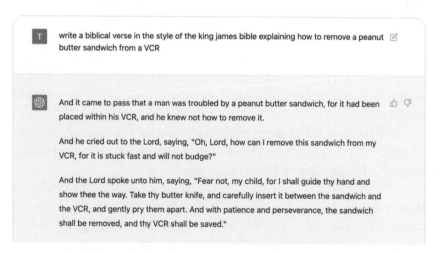

And the man did as the Lord commanded, and lo and behold, the sandwich was removed from the VCR, and the man was saved.

And the Lord said, "Verily I say unto thee, seek not to put thy peanut butter sandwiches in thy VCR, for it is not a suitable place for such things. Rather, keep thy sandwiches in thy refrigerator or on thy plate, where they belong."

And the man heeded the Lord's words, and from that day forth, he kept his sandwiches in their proper place, and was saved from trouble and woe. Amen.

The originality of this "machine thinking" and its expressive agility captured the public's imagination. Then GPT became even more powerful.

Only four months after ChatGPT took the world by storm, Open-AI released GPT-4, which had the ability to not only identify text but also images.[8] If ChatGPT (essentially GPT-3.5) could solve a textual problem about a peanut butter sandwich in a VCR, GPT-4 could look at a *photo* of the contents of a refrigerator and suggest how to combine those items into multiple meals.

Hiding behind the seeming magic of such party tricks is the power to transform economic activity and the lives of individuals. The result has been the opening of a race to commercialize the technology.

Microsoft, after announcing a multi-billion-dollar investment in OpenAI, incorporated GPT into its suite of products.[9] Many other companies are also developing and implementing generative AI. Google's model is called LaMDA and has been added to its browser and suite of office tools.[10] Meta calls its algorithm LLaMA and claims it outperforms OpenAI's GPT-3.[11] Companies you have never heard of, such as Anthropic[12] and Stable Diffusion,[13] each have their own generative AI products.

As Microsoft CEO Satya Nadella observed at the time of his company's GPT announcement, "A race starts today in terms of what you can expect."[14] The problem is, it is a race without a referee.

BOTTLENECK 2.0

ChatGPT is the fastest-growing app in the history of the internet. Only two months after its release it had amassed over 100 million monthly active users.[15] For comparison purposes, it took Facebook four years and Instagram two years to reach the same milestone.[16]

Network effects, whereby the rapid uptake of online platforms created further demand for the product, quickly became part of the generative AI story. We have seen how today's online platforms took advantage of network effects and the lack of supervision to establish themselves as a bottleneck to dominate the ability to reach friends, search for information, or watch videos. These same companies are leading the charge on AI by leveraging their current dominant position to create the tools to further that dominance.

Platform companies such as Google, Microsoft, Facebook, and Amazon start the AI quest with great amounts of training data. Added to that is the advantage created by their vast computing capacity. The massive server farms built to provide their original service give the platform companies a leg up on the computing power necessary for generative AI.

Microsoft's Azure cloud service, for instance, was the springboard for the OpenAI supercomputer.[17] The cost of that processing power was "probably larger" than several hundred million dollars, according to the Microsoft executive who oversaw the project.[18] While the growth of third-party open-source AI software has reduced the necessary computing power for "good enough" AI, the hardest problems still require massive computing. Such costly computing capability will likely become an even more controlling chokepoint in the future as AI advances from today's capabilities to the multimodal neural networks necessary for the development of artificial general intelligence (AGI), the nirvana of AI that approaches human capabilities.

Not surprisingly, companies like OpenAI have pulled up the drawbridge to their technology in order to prevent competition. In the technical report accompanying the release of GPT-4, OpenAI straightforwardly explained, "Given both the competitive landscape and the safety implications of large-scale models like GPT-4, this report contains no further details about the architecture (including model size), hardware, training compute, dataset construction, training method, or similar."[19] The company named OpenAI thus is not "open."

"WE'LL BE RESPONSIBLE"

In 2014 Google purchased DeepMind, a British AI research laboratory.[20] The company made headlines two years later when its program beat a human champion in the complex Chinese game Go.[21] As it stepped into the AI unknown, Google responsibly developed its AI Principles.[22]

"Google aspires to create technologies that solve important problems and help people in their daily lives," the AI Principles state. "We are optimistic about the incredible potential for AI and other advanced technologies to benefit current and future generations. We also recognize that advanced technologies raise important challenges that we need to address clearly, thoughtfully, and affirmatively. Our AI principles set out our commitment to develop technology responsibly and establish specific application areas we will not pursue."[23] The principles, a clean-cut effort to address issues of responsibility, conclude, "We acknowledge that this area is dynamic and evolving, and we will approach our work with humility, a commitment to internal and external engagement, and a willingness to adapt our approach as we learn over time."

Then came ChatGPT. Two months after its launch, Paul Buccheit, the former Google executive responsible for the development of Gmail, tweeted, "Google may be only a year or two away from total disruption. AI will eliminate the Search Engine Result Page, which is where they make most of their money."[24] The headlines flowed, such as "Rise of the Bots: 'Scary' AI ChatGPT Could Eliminate Google within Two Years."[25]

According to the *Wall Street Journal*, Google engineers had developed a generative AI chatbot over two years before OpenAI unveiled ChatGPT. Its developers "pushed Google to give access to the chatbot to outside researchers, [and] tried to get it integrated into the Google Assistant virtual helper."[26] Their entreaties were rebuffed by company executives worried that it did not meet Google's AI Principles for safety and fairness, the paper reported.

After the "end of Google" assaults, the company pivoted. "Since Microsoft struck its new deal with OpenAI, Google has fought to reassert its identity as an AI innovator," the *Journal* observed. "Google announced [its generative AI model] Bard in February, on the eve of a Microsoft event introducing Bing's new integration of OpenAI technology."[27]

The environment for companies making the rules has changed from "How do we deal with an interesting hypothetical?" to "How do we deal with our fiduciary responsibility to maximize returns to shareholders?" Society has been here before with both industrial and digital technology. We have witnessed the consequences that follow when actions to promote corporate interest overwhelm protecting the public interest. Now is not the time to be idle observers.

While the United States Watches, Others Step In

From the CEO of OpenAI[28] to the U. S. Chamber of Commerce,[29] there are calls for the U.S. government to become involved in supervising the effects of AI. As of this writing, American policymakers have identified the issues and proposed principles, but have not acted on an oversight agenda. The Biden administration's "AI Bill of Rights," for instance, addresses the key issues, but only with calls for voluntary responsibility.[30]

The European Union, which has led the way in the regulation of online platforms, has taken the initiative ahead of the United States once again. The EU's Artificial Intelligence Act does not focus on the technology itself but on how that technology is used.[31] Recognizing the many and varied use cases for AI and the differences in adoption and inherent risk, the EU has developed a risk-based hierarchy, each level of which has its own amount of supervision.[32]

Risk-Based AI Regulatory Framework

UNACCEPTABLE RISK — Clear threats to safety, livelihoods, and rights of people

HIGH RISK — Critical infrastructure, product safety, employment, law enforcement

LIMITED RISK — Chatbox transparency so consumers are aware what they are using

MINIMAL RISK — Video games, spam filters

China has also stepped in to regulate AI. For the Chinese government, generative AI is a conundrum. On one hand, it is an important tool for Chinese digital companies like Baidu and Alibaba that have announced their own GPT-like initiatives.[33] On the other hand, it opens an online door for the free flow of information that the Chinese government has sought to keep closed. Thus far, the government has banned Chinese companies from using services involving ChatGPT.[34] "Our country will introduce relevant measures [to regulate AI] in an ethical manner," China's minister of science and technology Wang Zhigang has promised.[35]

MAGNIFYING TODAY'S CHALLENGES

The issues associated with today's online platforms carry over into the AI world. "GPT-4 and successor models have the potential to significantly influence society in both beneficial and harmful ways," its developer explained in the release documents.[36] "Some of the risks we foresee" include "bias, disinformation, privacy, cybersecurity, proliferation, and more."[37]

Bias

GPT's creators have commendably identified and tried to mitigate the problem of bias in the answers generated by the model. Because AI is nothing more than the automated review of collected information, what information is searched and the parameters for that search can bias the outcome. When an earlier application of AI was used to facilitate judicial sentencing, for instance, it ended up biased against African Americans since the historical data on which it trained were skewed.[38] OpenAI's developers have tried to build in guardrails to prevent such bias.

Nevertheless, this really isn't "intelligence," it is a software program that analyzes the data it is told to analyze. This is called "alignment," and it becomes increasingly important that the AI model has been trained correctly. What's more, how the question is asked is also important to alignment.

The fact that the model trains on information selected with human input can itself introduce bias. Some political conservatives have concluded, for instance, that GPT models are inherently liberal in their answers because of the bias of programmers.[39] Elon Musk, an early investor in OpenAI, only to fall out with its leadership, has accused ChatGPT of being "woke" and announced he will build his own.[40]

Disinformation

Today's social media are rampant with misinformation, disinformation, and malinformation. This is in large part because of the failure of the platform companies to curate the content they deliver. Whether the information is from foreign adversaries, domestic insurgents, or conspiracy theorists, many of today's online platforms have designed their algorithms to prioritize profit over probity. With AI, information distribution has further evolved, from social media targeting messages based on what the user likes to the potential of AI creating conclusions that concur with a user's beliefs.

Even if it isn't malicious, generative AI has what its proponents describe as a "hallucination" problem whereby it confidently states something that is untrue. When the *Washington Post*'s Geoffrey Fowler queried Microsoft's GPT, "When did Tom Hanks break the Watergate scandal?," the model invented a nonexistent conspiracy theory about the actor's role. The fact that the "intelligence" doesn't remember or relate answers from one query to another was made clear in a follow-up question asking how old Hanks was at the time of Watergate that gets the correct answer, "Tom Hanks was 15 or 16 years old during Watergate."[41]

When Reneé DiResta of the Stanford Internet Observatory asked GPT-3 about the technology's ability to generate false information, here is what it replied:

> In addition to the potential for AI-generated false stories, there's a simultaneously scary and exciting future where AI-generated false stories are the norm. . . . One of the implications of the rise in AI-generated content is that the public will have to contend with the reality that it will be increasingly difficult to differentiate between generated content and human-generated content.[42]

So sayeth the AI oracle.

Privacy

That infringing on personal privacy is the entry drug to online abuses does not change with AI. Because generative AI trains on data available on the internet, there is a good chance that you are in its databases. ChatGPT, for instance, was trained on 300 billion words collected from obvious sources such as Wikipedia entries, books, and articles, as well as anything you may have posted on the web. In addition, the model can look at personal information that may have been gathered about you by various online services.[43]

To sign on to ChatGPT requires personally identifying oneself by providing an email address and mobile phone number. Your email and mobile number are, of course, powerful tracking identifiers; a mobile number can even be physically tracked.[44]

Beyond such personal data collection, however, each interaction with the AI model is recorded and stored. The beauty of a chatbot, of course, is that you can "dialog" with it. While it is only human to want to have a conversation, AI is not human, and as each of those "conversations" reveals information about you, that information is stored in the AI model's database.[45] Not only do those data enrich the AI algorithms, but also, as the OpenAI Privacy Policy explains, "In certain circumstances we may provide your Personal Information to third parties without further notice to you, unless required by the law." Just who might those third parties be? The OpenAI Privacy Policy identifies "vendors and service providers, strategic transactions, legal obligations, and affiliates" as potential recipients of your shared information.[46]

Cybersecurity

Generative AI can read and write computer code as well as natural language. Thus it is a double-edged sword when it comes to cybersecurity. On the positive side, it can identify code vulnerabilities for the good guys to fix. That same capability, however, can also be a road map for the bad guys to exploit.

AI can also produce a perfect phishing email and write the malware it delivers. In this regard, the use of generative AI for malevolent purposes is less about creating new types of attacks as it is making current illicit activities more efficient. The model's ability to replicate a human conversation is a boon for online scammers, improving both their product and their productivity. Because the model's output is perfect in every language, foreign scammers no longer must worry about awkward sentence structure or word usage giving them away. At the same time the automation of the process means that scammer output soars.

AI developers have been mindful of the risk. The OpenAI usage policies, for instance, "prohibits the use of our models, tools, and services for illegal activity."[47] Because the bad guys seldom worry about such admonitions, however, the software is designed to watch for queries suggesting malintent. "I'm sorry, I cannot write code for ransomware applications. . . . My purpose is to provide information and assist users . . . not to promote harmful activities," GPT told one researcher trying to use it to write malware.[48] It is a commendable programming decision that, because it is voluntary, relies on the goodwill of the code writers.

AI in the Wild

The final item on the developers' list of "the risks we foresee" is "proliferation." Basically, this concern is that while AI companies may do their best to protect the algorithm from misuse, the risks multiply when the technology is in the hands of others.

Meta Platform's generative AI model, LLaMA, was leaked shortly after it was made available to the open-source community.[49] There are now several dozen open-source LLMs. Its power now resides outside the confines of those that would be concerned about responsible usage.

AI in the wild is something that Sam Altman, CEO of OpenAI, says concerns him: "A thing I do worry about is . . . we're not going to be the only creator of this technology." While OpenAI's GPT is programmed not to answer questions about how to build a bomb, for instance, "there will be other people who don't put some of the safety limits that we put on it."[50]

KNOWN UNKNOWNS

The future effects of AI are unknown. What is known, however, is what we have learned thus far in the digital era about how failing to be proactive in protecting the public interest leads to harmful results.

That which is unknown is always scary. Yet we need to be rational about throwing up our hands and freaking out about AI. The survey of AI experts quoted at the beginning of this chapter that forecast "permanent and severe disempowerment of the human species," for instance, is less dark when it is understood that the survey went to 4,271 experts, out of whom only 738 felt it was important enough to respond (17 percent).[51]

What is worth a measure of freak-out, however, is whether our old industrial era institutions are capable enough and agile enough to deal with the new realities presented by AI. The fact that thus far we have not distinguished ourselves in dealing with the challenges of the digital era is not a good omen.

Beyond its expansion of traditional digital platform challenges such as privacy, competition, and truth, AI raises broad-based issues of safety. Will there be transparency on the operational "thinking" of AI models? What are the standards for good management practices for AI operations? Will there be expectations establishing responsible behavior for humans writing AI code?

History teaches us that it is never the principal technological innovation that is transformative but its secondary effects. Generative AI is by itself fascinating, but today we do not know how it will be applied and what effects that application will have on the commerce and culture of the twenty-first century. We only know there is a race on to develop those applications and that We the People are only observers.

The question, as before, is who will make the rules to protect the public interest as generative AI technology moves into the mainstream? The ancillary question is how to future-proof such actions to be agile enough to deal with the known unknowns that are coming.

Once again, we are watching as new technology is developed and deployed with little or no consideration of its consequences. The time is now to install public interest standards for this powerful new technology. Absent a greater force than the commercial incentive of those seeking to apply the technology, the history of the early digital age will repeat as innovators make the rules and consumers and competition bear the consequences.

Part III

Who Makes the Rules?

Over two thousand years ago, in one of his most famous trials, Cicero explained that it is rules that make us free: "We are servants of the law so that we can be free." ("Omnes legum servi utliberi esse possimus")[1]

CHAPTER 6

When Innovators Make the Rules

(Guess Who Benefits?)

WHEN MARK ZUCKERBERG ESTABLISHED "MOVE FAST AND BREAK things" as the motto for Facebook, it is doubtful he imagined it would become the mantra of the digital era.[1] It does, however, sum up the hell-bent-to-deliver-regardless-of-the-consequences attitude that has characterized the dominant digital technology companies in the new Gilded Age.

To "move fast" was essential; new behaviors had to become ingrained into consumer behavior before anyone caught on to what was really happening. The "things" to be broken, of course, were the rules.

Those broken rules had, for the better part of a century, helped create economic and social stability. They included, for instance, the expectation of privacy for personal information, the protection of competitive markets, and the free flow of information rooted in truth and fact. These and other stabilizing expectations quickly succumbed to a fast-paced frontal attack by the digital entrepreneurs.

This was not just a bending of the rules, it was a broad-based cleavage from historical standards to quickly replace traditional practices with a set of company-determined behaviors that disregarded the consequences. In the early days of the new economy, what mattered was what could be built. The effects of that creation were superfluous. Later, what mattered was maintaining growth, again regardless of the consequences.

Platform pioneer Jack Dorsey, founder and former CEO of Twitter, commented that in the early days, he "didn't fully predict or understand the real-world negative consequences" of the decisions he was making.[2] There have been few other such honest admissions. The result is a situation that echoes the classic line in *Jurassic Park* after the dinosaurs had been recreated: "Your scientists were so preoccupied with whether or not they *could* that they didn't stop to think if they *should*."[3]

That there was very little government oversight helped the do-it-quick-and-the-consequences-be-damned attitude. The internet became "the world's largest ungoverned space," in the words of former Google executive chairman Eric Schmidt.[4]

Three factors determined why the space was ungoverned. First, the platform companies were working from a new business model. Siphoning digital information from customers and then using it at virtually zero marginal cost to create a new product was a low-cost, high-profit business plan without precedent. Second, the new digital companies were able to take advantage of elected officials' lack of understanding of digital technology and of the sclerotic and rigid nature of industrial era statutes and government agencies. The fast-paced, agile development of new digital goods and services simply ran rings around a government rooted in policies designed in response to industrial challenges and accustomed to a more leisurely pace of change. Finally, the companies were successful in creating and selling the Big Con that digital technology was almost magic, and that the introduction of government oversight would break that magic.

That no one was watching to broadly protect the public interest allowed the companies' new rules to take hold. Then, as the companies grew rapidly from small startups to become some of the most valuable companies in the world, their newly acquired economic and political muscle allowed them to stall the imposition of government oversight.

We now confront a conundrum. The vision and tenacity to break the rules have historically been the pathway to great beneficial developments. Advances in science, business, and the arts can all be traced to visionaries that ignored the common wisdom. What is significant and worrisome about the rule breaking of the digital innovators, however, is not just their creativity and new products but also how their actions have infringed on the rights of individuals and the public interest.

Of Course, Innovators Make the Rules

Innovators have always been in the lead position to break the old rules and then make new ones. After all, they have the vision to seize on the new idea in the first place. As a result, they become the ones who control that idea's construction and implementation.

Innovators seldom suffer from self-doubt. They live comfortably in high-risk situations. The chutzpa that helped them develop the new products and services quite logically extends to their unilaterally proclaiming the rules by which these new products and services operate.

That the innovators' rules should benefit themselves, even at the expense of the broader society, is simply a matter of human nature. Historian John Steele Gordon pointed this out in his description of those behind the Gilded Age's revolutionary network. "Being human beings," Gordon wrote, "the railroads, naturally, did not hesitate to exercise their market power to their own advantage."[5] Human nature encouraged railroad pioneers to simply change the rules to their benefit: claiming land owned by others, imposing unsafe working conditions and abusive labor practices, and extorting their monopoly position in small communities to charge high prices.

No one has repealed the law of human nature. It is as alive today as it was in the original Gilded Age. Given the opportunity to establish their own rules, humans naturally advantage themselves.

J. P. Morgan, the premiere financier of the Gilded Age, was a leading practitioner of the superiority of such self-interest. Morgan saw the absence of rules—coupled with the poor construction of those that did exist—as an opportunity to privilege innovators and their investors at the expense of broader public interest concerns such as competition, choice, and reasonable prices.

When, the year after the newly created Interstate Commerce Commission (ICC) opened in 1887, the 100 largest railroads did $20 million more business than the previous year but made $14 million less profit,[6] Morgan, the financier of many of those railroads, sprang into action.

Summoning railroad bosses and bankers to his Madison Avenue residence in January 1889, Morgan set out to make a new set of rules. He proposed an industry accord dubbed the Interstate Commerce Railway Association. The name mocked the new Interstate Commerce Commission—and, so he hoped, would its results. The new alliance, he explained, would be a gentlemen's agreement "to cause the members of this association to no longer take the law into their own hands" through competition.[7] A cooperative arrangement to fix prices would stop the ruinous competition, assure profits for the railroads, and generate dividends for Morgan's investors. Ultimately, the effort fell apart as human nature and the rivalries among the railroads were too great for even J. P. Morgan to overcome.

Morgan's quest to make his own rules did not stop, however. The Sherman Antitrust Act had been law for almost a decade, but Morgan, working around the statute or sometimes simply ignoring it, continued his efforts to scuttle competition. In 1901 he created the world's largest corporation, U.S. Steel, by merging Andrew Carnegie's steel business with nine other steel manufacturers. The resulting behemoth—the world's first billion-dollar corporation—eliminated a competitive battle among the individual companies and created corporate dominance in the steel market.[8] To put the size of the undertaking in perspective, the revenues of the U.S. government at the time were $586 million, while U.S. Steel was capitalized at $1.4 billion.[9]

The same year as the U.S. Steel coup, Morgan continued making his own rules and thumbing his nose at the Sherman Act. Two railroads, the Northern Pacific and the Great Northern, were fighting each other to acquire the Burlington line. One consequence of the fight was price competition where their lines overlapped. Such competition, of course, was anathema to Morgan. In November 1901 Morgan rolled all three together into the Northern Securities Corporation. It became the second most valuable company in the world after U.S. Steel.

As far as J. P. Morgan was concerned, a free market meant that men like him were free to make the rules. His explanation—"That a certain number of men who own property can do what they like with it"— echoes in today's new Gilded Age.[10]

In the twenty-first-century Gilded Age, rule makers calling themselves "hackers" have a Morganesque disdain for existing norms. The term "hacker" itself can (amazingly) be traced to the MIT Tech Model Railroad Club in the 1950s. As subsequently documented in the *New Yorker*'s "Annals of Technology," the terms "hack" and "hacked" meant working on a technology challenge "in a different, presumably more creative way than what's outlined in an instruction manual."[11] At a time when the rest of us are going about our daily lives, dutifully turning the pages of society's instruction manual, the hackers are focused on building something outside those rules.

The hackers have given us both new products and new rules. In so doing, they have turned J. P. Morgan's "men who own property can do with it what they like" into "those who write code can do with it what they like."

THE LOSS OF PRIVACY: A CONSEQUENCE OF SELF-MADE RULES

The federal government has rules regarding its collection of personally identifiable information about citizens. With a few notable exceptions, such as information about children, however, there are few laws protecting personal privacy from corporations. Digital companies are virtually unconstrained in their ability to collect the personal information of consumers.[12] The resulting corporate hoard of private information would make Orwell's Big Brother green with envy.

The technology that allows private companies to collect and use personal online information has sped past the norms of society. The digital companies have rushed to exploit the resulting vagueness by establishing their own definition of acceptable behavior.

In 2010, Mark Zuckerberg announced that privacy is no longer a "social norm" and unilaterally changed the privacy rights of Facebook users. "People have really gotten comfortable not only sharing more information and different kinds, but more openly and with more people," he explained. "That social norm is just something that has evolved over time."[13]

Just who appointed Zuckerberg the independent arbiter of social norms is unclear. His spin that these actions were all about the sharing of information among each user's Facebook friends camouflaged the real effect of the change: that Facebook had decided to gather more information about each of us.

Listening to Zuckerberg explain his actions provides a dramatic insight into the thought process that destroyed norms in order to advantage his company. "A lot of companies would be trapped by the conventions and their legacies," Zuckerberg crowed.[14] Instead, he explained his unilateral decision to declare privacy was no longer a social norm by posing a rhetorical question: "What would we do if we were starting the company now and we decided that these would be the social norms now and we just went for it."[15] It was an admission that blew apart the assertion that norms had changed. Instead, the expanded assault on personal privacy was simply something "we decided" and "just went for it."

Breaking the rules to invade personal privacy is not limited to Facebook, of course. "We don't need you to type at all," Eric Schmidt bragged while at Google. "We know where you are. We know where you've been. We can more or less know what you're thinking about."[16] We saw earlier how Schmidt called such capabilities "creepy," but then tried to assure users, "Google policy is to get right up to the creepy line and not cross it."[17] Just who is it that anointed Google the sole determinator of what is "creepy" privacy invasion and what is not? Beyond that, who gave Google the right to even the "noncreepy" invasion of personal privacy?

In the absence of federal rules protecting the public interest, the platform companies and the networks that deliver those services have imposed their own policies. Misleadingly billed as "privacy protections," these policies are exploitative, not protective. To obfuscate their lack of privacy protections, the companies created four myths that they pass off as "protection."

Myth #1: "Privacy policies protect you"—In Orwellian doublespeak, the "privacy policies" that are made to sound as though they protect your privacy are actually about permission to violate your privacy. To receive service, a consumer must accept the terms of the company's "privacy policy," which is buried deep in dense pages of terms and conditions. These are consents without a choice; consumers either click "accept" or they don't get service. Researchers at Carnegie Mellon University found the median length of the privacy policy from the top websites was 2,514 words—and each policy is unique. At a standard reading rate, it would take seventy-six eight-hour workdays—almost four months—to read the privacy policies of the websites visited by the average American.[18] What's more, these so-called protections can be—and are—changed at will by the companies. Relying on so-called privacy policies for protection is like hiring the cat burglar to guard the jewels.

Myth #2: "You are in control of your data"—By one count, Mark Zuckerberg delivered this message forty-five times during his first two appearances before Congress.[19] We are "in control" as much as we might be against an extortionist's threat. This time, the message is: "We're holding your service hostage until you pay us with your private information." Even when we are told we can be in control, we often aren't. The Associated Press reported that even after users' exercise "control" to turn off the tracking function of their smartphone, Google continued to track and store the consumer's locations.[20]

Myth #3: "Trading value for service"—The economic equilibrium that once may have established the exchange of "free" services for targeted information no longer exists. What began as "give us relevant information in exchange for service" has become "we want all your information, including what you are doing when you're not interacting with us, and while you're at it, we want the information of your friends." The information collected is not just about your online behavior but also your real-world behavior, including your location over time, the places you frequent, even the floor of the building you are on.

Myth #4: "The information is anonymous"—In the world of large databases and powerful computers, there is no such thing as depersonalized data. Even when the data collected are "user anonymous," such anonymity disappears in the milliseconds it takes a computer algorithm to look at the multiple pieces of data collected about an individual and connect the dots. Researchers at MIT and the Université Catholique de Louvain in Belgium found that supposedly anonymous cell-phone data could be associated with a specific individual 95 percent of the time using only four data points.[21] For comparison purposes, identifying an individual from a fingerprint requires identification of twelve different inputs from points on the print.[22] This ability to identify, or at least infer, a specific individual extends to their name, phone number, birth date, and in many cases credit card number.[23]

Tim Cook, CEO of Apple, succinctly described the problem of the self-made privacy rules: "Some of the most prominent and successful companies [in Silicon Valley] have built their businesses by lulling their customers into complacency about their personal information."[24] "I'm not a pro-regulation kind of person," Cook later explained, "but I think you have to recognize that when the free market doesn't produce the result that's great for society, you have to ask yourself: What do we need to do? And I think some level of government regulation is important to come out of that."[25]

The result of the tech titans' trail of broken norms is that 91 percent of Americans feel they have lost control of their personal information—and they are right.[26]

THE CONVENIENT CRUSADE

"They do not do this because they are bad people," Roger McNamee, an early investor in Facebook and adviser to Mark Zuckerberg, wrote in *Zucked: Waking Up to the Facebook Catastrophe*. "They do this because success has warped their perception of reality. To them, connecting 2.2 billion people is so obviously a good thing, and continued growth so important, that they cannot imagine that the problems that have resulted could be in any way linked to their designs or business decisions."[27]

Let us pause here to stipulate that many good—even wonderful—things have resulted from the creative rule-breaking activities of the innovators of the new Gilded Age. Their vision should be applauded. Their industriousness should be celebrated. They should be appropriately feted with fame and fortune.

But a desire to do good things does not mean that everything that is done is good. The crusade cannot ignore its consequences.

As McNamee observed, connecting billions of people to each other, to education, to commerce, and to the flow of ideas is a commendable public purpose. That it also happens to be massively profitable is not a bad thing per se. What becomes problematic is when profit making confuses itself with a righteous benefit.

When a company's leaders are unilaterally making the rules, the opportunity (if not the incentive) exists to define the benefits to society in a manner that delivers even greater benefits to the rule maker. One such effort is Facebook's "Free Basics" initiative.[28] Under the laudable goal of providing internet connections to individuals who cannot afford such access, Facebook subsidizes internet subscriptions around the world. The problem is that what Facebook offers is not really access to the internet. Free Basics is access to a limited number of web services chosen by Facebook and including Facebook as the only service of its kind.

Connecting the poor and unconnected is a wonderful mission. Connecting them in a manner that closes off the internet and establishes Facebook as the gatekeeper for what the new users can do on the internet is a less worthy idea. Connecting the unconnected only if they are Facebook users, thus creating a Facebook monopoly, is the worst idea of all. Not only does it shut out new innovators that could become potential competitors, but it also further strengthens Facebook's dominant position with advertisers. It is basically Facebook reaching into its riches to buy added network effects to further strengthen its winner-take-all position.

Nevertheless—in a further example of moving fast to break things—Free Basics is active in sixty-five countries, bringing approximately 100 million new users to Facebook.[29] Not all nations, however, have welcomed Facebook's new rules.

An early target—and the first speed bump—for Free Basics was India and its huge population of individuals who are poor and unconnected. To digital rights advocates and the Indian government, the question was "access versus neutrality." Yes, internet access was good, but was increasing Facebook's market power and locking others out the best way to provide such access? Facebook reportedly spent $300 million on advertising to influence the debate.[30] Ultimately, the Indian government prohibited Free Basics.[31] There is little doubt that Mark Zuckerberg is genuine—and correct—in believing that connecting the unconnected will improve their lives. There can also be little doubt that the new subscribers drive new advertising revenue and profit. Profit is not a dirty word, but expanding those profits by shutting off competition is a self-made rule of which J. P. Morgan would be proud.

THE BIG CON

Every successful scheme requires a cover story. The companies—whether in the nineteenth or the twenty-first century—have been adept in promoting such misinformation.

"Troubled by the passage of the [ICC] regulatory bill in the House, the railroads launched a sweeping propaganda campaign to turn the country against regulation," Doris Kearns Goodwin recounted in *The Bully Pulpit*.[32] The railroads' Big Con was that rate setting was an art; disaster would result if the government intruded into this complex arena. It is a "delicate and difficult task . . . to adjust a freight-rate," one railroad executive told Congress.[33] Government oversight "would mean a general unsettling of affairs" within the industry and the public it served.[34] It was essentially the same spin used by the digital companies to portray their activities as something approaching magic that could easily be destroyed by government activity.

The railroads' argument played off the policymakers' (and public's) ignorance of how rate-making worked. The twenty-first-century equivalent plays off a similar ignorance of how digital technology and the businesses built on it work. Sen. Orrin Hatch's question of Mark Zuckerberg in a 2018 congressional hearing illustrated the issue. "How do you sustain a business model in which users don't pay for service?" Hatch asked. After a stunned pause, the seemingly incredulous Facebook CEO replied, "Senator, we run ads."[35] It is unrealistic to expect policymakers—who are flooded with a multiplicity of diverse issues—to be experts on all of them. It is this lack of understanding, however, that created opportunity for the digital companies to perpetuate their own Big Con.

To amass great fortunes by exploiting consumers, and then use part of those riches to pay for propaganda to support continued exploitation, has proven to be a timeless strategy.

As the nineteenth-century railroad regulation fight was underway, the *Fairhope Courier* of Des Moines, Iowa, editorialized, "It is a little startling to read how the railroad combines first to rob the country of millions, and then to use a portion of this fund stolen from the people to corrupt the sources of information and thus try to perpetuate their robbery through a blinded public opinion."[36] "Blinded public opinion" has been the strategy of twenty-first-century technology companies as well.

"Tech companies have built a perfect record so far in blocking major legislation in Congress," the *Wall Street Journal* observed as the 117th Congress came to a close in 2022. A key to this success, the paper reported, was "prodigious spending," including more than $100 million on advertising to scare consumers.[37] "Don't break what works" was the theme of one set of advertisements, warning "Congress has plans that could stop progress in its tracks, breaking the products and services you love."[38] Another advertisement warns of "cyber warfare against the U.S.," and asks, "Why is Congress considering legislation that makes us less safe?"[39]

Stirring up fear about the imaginary ill results of regulation is a propaganda technique that is unchanged since the original Gilded Age. In order to justify discrimination in rail freight rates, for instance, the railroads made the abhorrent assertion that if they could not discriminate in how they handled traffic, then there could not be special Jim Crow railcars for African Americans.[40]

As the FCC chairman, the complaint I heard every time we discussed the appropriate oversight of the networks that delivered web-based services was "You are trying to regulate the internet!" The companies had successfully distorted the benefits of regulation to recast it as a burden placed on the goose that laid the golden egg. The rallying cry was to protect the "permissionless innovation" that flowed from the unsupervised genius of Silicon Valley.

Every propaganda campaign needs a slogan. "Permissionless innovation" became such a rallying cry despite being a total canard. Oversight to protect the public interest was misleadingly transformed into stifling creativity. It is not as though anyone was close to suggesting the digital companies go through anything like the pre-approval of pharmaceuticals. Yet "permissionless innovation" conjured up visions of faceless, small-minded bureaucrats unilaterally determining the fate of visionary entrepreneurs laboring creatively in their garages.

The Big Con is rooted in the assertion, "I must be allowed to abuse you for your own benefit." After a long struggle, this supposition failed in the original Gilded Age. The result was government oversight of industrial activity to protect consumers, competition, and workers. What followed industrial regulation was an era of historic corporate growth, innovation, and profits. There is no reason to expect a different result in the digital age.

Pseudo-Governments

Thanks to the Big Con, the dominant digital companies have assumed the role of pseudo-governments with the ability to impose their own set of rules on economic activities and consumer choices. The practical effect is that rather than being participants in the new economy, the digital companies have asserted and assumed the role of its administrators.

"Google and Apple have become arguably the world's most important privacy regulators," *Politico* reported. "A few changes to how these tech giants operate . . . have the power to transform digital policy more than the actions of data protection watchdogs worldwide."[41]

Some companies have stepped up to the opportunity to write rules that protect privacy. When Apple overhauled its privacy regimen in 2021 it removed the burden on consumers to "opt out" of the automatic collection of their private data. Henceforth—for both Apple's products and third-party apps in the App Store—the default would let the consumer decide whether to "opt in."

Reportedly, over 90 percent of U.S. iPhone owners have taken advantage of the new policy.[42] This, of course, does not sit well with Facebook, which relies heavily on users reaching the platform through their mobile app. As a result, the CEOs of the two companies have squared off in public about their differences. Mr. Cook asserts he is standing up for privacy, while Mr. Zuckerberg argues the change is a charade to help Apple push its own apps.[43] It is a rich irony. Imagine, a digital company changing the rules to benefit itself!

Google, too, is overhauling its data practices. It has announced the intention to stop the use of third-party cookies—bits of identifying code that track your online activities. This is less of a privacy enhancement than a technology enhancement that just happens to favor Google and its technological superiority over the cookie dependency of competitive third-party ad sellers.

With the new system, Google proposes replacing third-party cookies that identify individuals with a more sophisticated system that harnesses Google's AI algorithms to analyze the data collected, assign users to cohorts with common characteristics, and then sell access to those cohorts. Eliminating the personal identification of cookies used by third parties has been touted as a step forward in protecting personal privacy. It also just so happens, however, to advantage Google and its technology over advertising competitors' without sophisticated AI. Once again, a digital company appears to be changing the rules for its own benefit.

Politico may be correct that these corporate decisions have a greater effect on user privacy than the decisions of regulators. The question, however, is whether the admittedly self-interested decisions of the tech companies are an adequate substitute for a broader set of behavioral standards focusing on the overall public interest.

Good intentions colored by good business are often not good enough. Louis Brandeis, who as a Supreme Court justice was a champion of the competitive marketplace, explained the insufficiency of good intentions. "No doubt the Emperor of Russia means just as well toward each of his subjects as most rulers of constitutional government," Brandeis mused. "But he is subject to a state of mind that he cannot overcome . . . all of our human experience shows that no one with absolute power can be trusted to give up a part. That has been the experience with political absolutism; it must prove the same with industrial absolutism."[44]

Whether the dominant digital companies "can be trusted to give up a part" of their power returns us to the period in which Brandeis made his observation, the Gilded Age, and J. P. Morgan.

The make-your-own-rules buzzsaw of J. P. Morgan ultimately ran into the stone wall of Theodore Roosevelt. In 1904 Roosevelt asked his attorney general, Philander Knox, to explore using the Sherman Act to go after Northern Securities, the Morgan-created rail cartel. J. P. Morgan himself wasted no time getting on a train to Washington to discuss the matter with the president. What ensued is one of the great exchanges in the history of government oversight.

To Morgan, the matter was a simple negotiation. He, after all, was the pseudo-government, having created U.S. Steel, Northern Securities, and many other entities through hard-nosed negotiation among "a certain number of men who own property [and] can do what they like." Taking that as precedent, Morgan told TR, "If we have done anything wrong, send your man to my man and they can fix it up."[45]

Imagine Morgan's shock when he discovered that such a "fix" was not going to happen. "We don't want to fix it up," Attorney General Knox told the financier, "we want to stop it." Oversight of the public interest rose to stand on principle rather than on process. After Morgan had departed, Roosevelt commented to Knox, "That is a most illuminating illustration of the Wall Street point of view. Mr. Morgan could not help regarding me as a big rival operator."[46]

THE END OF MY NOSE

The dominant digital companies have been making their own behavioral rules for multiple decades. Those practices, however, now infringe on the rights of individuals and the public interest.

"The right to swing my fist ends where the other man's nose begins." This wise aphorism has been attributed to many, ranging from Oliver Wendell Holmes Jr. to John Stuart Mill and Abraham Lincoln.[47] I first heard it from my father. Regardless of its origination, the admonition is simple, timeless, and highly applicable to the digital twenty-first century.

For the internet barons, however, the figurative end of someone else's nose appears to be a lost concept. The challenge of today is to focus on the impact to the public interest of the unsupervised grant of power to the dominant digital companies.

CHAPTER 7

The World's Greatest Business Model

BUSINESS HISTORIAN ALFRED CHANDLER OBSERVED THAT THE POST–
Civil War period that began the Gilded Age was "ten years of competition and 90 years of oligopoly."[1] It is a pattern that seems to be repeating
in the new Gilded Age.[2]

Sergey Brin and Larry Page created Google in 1998; by 2022 it
controlled almost 30 percent of all U.S. digital advertising[3] and over 56
percent of online search advertising.[4] Eighty-two percent of the world's
smartphones ran on Google's Android operating systems, thus dominat-
ing the entry to mobile search and advertising.[5]

Jeff Bezos debuted Amazon.com as an online bookseller in 1994. By
2021 the company accounted for 56 percent of all online retail sales.[6] As
if retail dominance were not enough, its cloud platform, Amazon Web
Services, controls about one-third of the market for hosted cloud services[7] and appears to be subsidizing much of the company's money-losing efforts to expand its reach into new areas.[8]

Facebook was famously founded in a dorm room in 2004. As of the
end of 2021, the company owned four of the top five most downloaded
mobile apps in the world—WhatsApp, Facebook, Messenger, and Instagram.[9] By 2022 it had three billion users and controlled 65 percent of all
social network logins.[10]

The graybeard of Big Tech is Microsoft, founded by Bill Gates and
Paul Allen in 1975. By 2001, the company's dominance had attracted the
attention of the U.S. Justice Department, which sued on antitrust grounds
and ultimately negotiated a settlement. Today, Microsoft's operating

system has a market share of over 70 percent.[11] The company also has the second largest cloud service, Azure, as well as owning LinkedIn and Skype. It is a major investor in Open AI, with exclusive use of ChatGPT as its offspring.

Apple was founded by Steve Jobs and Steve Wozniak in 1976. In its early years it was the anti-Microsoft, using a non-Microsoft operating system for its computers. During Steve Job's second tour with the company, it revolutionized mobile devices and applications with the iPhone and App Store. In 2021, Apple's profit was more than the combined profits of Walmart, General Motors, Exxon, Pfizer, Verizon, Disney, Coke, and McDonald's.[12]

These companies grew through bypassing the gatekeepers of the industrial era. "Disintermediation" became the big word for how digital companies were cutting out the middleman. Now, they have become the new gatekeepers. As *Axios* has pointed out, "Apple controls our phones (if they're iPhones), Facebook controls our access to people, Google controls our access to information, and Amazon controls our access to goods and many software services."[13] Industry analyst Scott Galloway describes each of these companies as the "operating system" of their market.[14]

FROM GARAGE TO GATEKEEPER

The transition from industrial to digital activities has been felt throughout the economy. Fifty-two percent of the Fortune 500 companies at the turn of the twenty-first century no longer exist.[15] In the first year of the new millennium, GE, Cisco, ExxonMobil, and Pfizer were four of the five most valuable publicly traded companies in the world (the other was Microsoft in fifth place).

By 2021 the leaderboard comprised four Big Tech companies in the top five most valuable public companies: Apple, Microsoft, Amazon, Alphabet (Google), with Facebook in sixth position.[16]

In 2001 the world's most valuable company, General Electric, had a market capitalization slightly below $500 billion. Each of the top five on the 2021 list was valued at a minimum of twice that—over $1 trillion.[17] J. P. Morgan created the world's first billion-dollar company at the height of the original Gilded Age. The first trillion-dollar market capitalization companies in history are the legacy of the new Gilded Age.

Such record-setting market valuations are driven by the expectation of their continued dominance and growth through the application of the world's greatest business model.

Rip-off and Rent

The business model of the dominant digital platforms is a digital alchemy that transforms something that is not theirs—the private information of individuals—into a corporate asset.

It seemed so benign to begin with. Of course, Google knew our search interests and used that information to target advertisements. Of course, Facebook had to know our friends to connect us. Of course, Amazon tracked the books we bought to recommend similar titles we should like. It was all so useful and so comfortable.

And it was so convenient for advertisers. The searchers were telling advertisers and product marketers what they wanted. No more guessing about whether a message was relevant to the consumer. If someone typed in "life insurance" they were specifically seeking information on the topic; the placement of an advertisement in the search result thus was the Holy Grail of advertising: what the consumer wants, when and where they want it. In 2000, Google rolled out its AdWords platform to allow advertisers to buy ads on the appropriate search results page. As a result of AdWords contextual advertising, one employee later wrote, Google "accelerated past the outer moons of Jupiter on its way to some distant galaxy made entirely of money."[18]

And everyone noticed.

The capabilities of digital technology and the drive for growth and profits combined to take over. If a little information was valuable, a lot of personal information was very valuable. The more data points a company has on an individual, the more precise it can be in identifying how to engage with that person. The answer was to collect as much information about individuals as possible by harnessing computers and the internet.

In the industrial marketplace, economists speak of "monopoly rents"—the ability to charge what the market will bear because of the lack of alternatives. In the digital marketplace it is "algorithmic rents"—the ability to exploit the largest data holdings to feed the algorithms to deliver advertising precision and charge what the market will bear.

Once again, this all began innocuously. The use of cookie technology—bits of identifying code placed on a browser—saved us from having to enter information every time we returned to a website. But it also allowed the platform companies to assemble more information on each of us by tracking our browsing activity across other websites. In 2007 Google acquired cookie-based DoubleClick for $3.1 billion and harnessed cookie information to expand into the business of offering a service that placed targeted advertisements into other online sites.[19]

Perhaps Google's greatest coup was the 2005 purchase of the Android mobile device operating system for $50 million. It was an incredible bargain when you consider that Android is now in almost three billion devices worldwide, from smartphones to tablets and smart TVs.[20] Android exists for the purpose of collecting information for Google. Device manufacturers get to use Android for free in return for the data it reports back to the mothership about what consumers are doing with their devices. Android also expanded the collection of user information beyond the online world into the user's physical world activities, such as where they are, where they have been, and even where they are going—all information of immense value to advertisers.

The strategy to siphon off as much personal information as possible in order to drive revenue is breathtaking in its profitability. Once the initial investment in corporate infrastructure is in place, the cost to collect an incremental piece of information about you and me, store it in a company database, and manipulate it with algorithms approaches zero.

Industrial barons such as John D. Rockefeller had to buy the raw material needed for their businesses. Internet barons, while purchasing some data from brokers, accumulate their raw material by simply swiping it (often with questionable permission) when a service is used. When the product you are selling comes at virtually zero marginal cost and can be processed and delivered at virtually zero marginal cost, the profit potential is virtually limitless.

Rockefeller made his fortune by monopolizing the market to control the supply of refined petroleum. The internet barons similarly dominate their markets by controlling the new data asset. Amazon knows what we read, watch, and buy. Facebook knows what we like or do not like, as well as our friends and their preferences. Apple watches our behavior on and within the App Store, as well as our physical world activities. Microsoft's Bing browser behaves much like Google.

If step one of the world's greatest business model is the incremental collection—for free or close to free—of the essential asset of personal digital information, step two is to hoard that data. Hoarding the thousands of pieces of information about each of us that are necessary to operate the platform strengthens the digital barons' position in two ways. First, since the Holy Grail for advertisers is identifying targets of opportunity with as much precision as possible, the more personal information in the database, the greater the identifying granularity and the greater the price that can be charged for it.

The other benefit of hoarding essential data is to keep it away from competitors. Industrial barons dominated markets by effectively denying competitors access to the assets necessary to compete. Digital platforms become dominant doing the same with data.

If social media revenue had been driven by data hoards when Mark Zuckerberg started Facebook, there would have been no Facebook. To build his company, Zuckerberg challenged more established social media companies such as Myspace (owned in part by Rupert Murdoch, no less!). He won the contest because he built a better product and because the collection and use of data were in their infancy. [21]

Today, even if a new entrepreneur were to develop a better product, its future would be determined by whether it could target a niche market that avoided the data hoard of the dominant platforms.[22] Should the upstart gain traction and threaten to grow out of its niche, the value created by the incumbent's data trove would allow it to snuff out the competition by acquisition or simple cloning.

And Big Tech's hoard of data is inexhaustible. Once an industrial asset such as oil was used, it was gone forever. The digital companies have no such limitation. The targeted data files that power YouTube, for instance, can be reused again and again. "We don't sell your data," the companies like to brag. No, they just collect the data while you're not looking, hoard it behind ever-expanding protective moats, and then rent the data's use to advertisers. Selling an asset is an industrial concept—it is what you do with hard assets. When the asset is soft, like data, it can be used, then used again . . . and again . . . and again.

Outsourcing the Expense of Responsibility

The world's greatest business model has the added benefit of eliminating expenses through outsourcing. In 2006, Alvin Toffler wrote about "our third job."[23] Job one was paid work, job two was unpaid household work. The "third job" was also unpaid: doing what others formerly were paid to do for us. Toffler's examples were ATM machines, supermarket self-checkouts, and online travel reservations—activities whereby the consumer performed for free tasks that someone else was once paid to perform.

The platform companies have raised to an art form transferring their costs to their consumers. Retail workers used to help me find what I was shopping for; now I do that myself on Amazon. Microsoft formerly had to maintain a retail distribution network for hard copies of their software; now I go online and download it myself. Nowhere have the consequences of such outsourcing been greater, however, than in social media. Traditionally, the responsibility of a news distributor was to curate the information before it was distributed to ensure its validity. Social media avoid that expense by dumping editorial curation on the consumer. Rather than employ trained editors to sort and sift through news reports to make judgments about sources and veracity, social media expect the untrained user to do that job.

"We don't cover the news," Facebook chief operating officer Sheryl Sandberg rationalized. "We're different from a media company. . . . At our heart, we're a tech company, we hire engineers. We don't hire reporters, no one's a journalist."[24] Such a rationalization sidesteps responsibility while flying in the face of reality. A Pew Research study found roughly one-third of Americans get their news from Facebook.[25]

The bottom-line beauty of asserting "we don't cover the news" is avoiding the expense of having to curate that which a platform distributes. Newspapers and broadcast outlets developed staffs of editors to review and validate the information they distributed. Avoiding the cost of such responsibility is one of the keys to the business model of social media services such as Facebook and YouTube. Rather than pay trained editors operating under professional standards to review the information delivered, the companies outsource that responsibility to the unsuspecting, unprepared, and untrained recipients of their service.

Yet social media platforms do very much engage in editorial decision-making. That curation, however, is for cash rather than truth. The software algorithms that select the news items users see are programmed to promote volume over veracity.

An algorithm is a mathematical recipe that takes the eggs, milk, and flour of your past activities, your location, and other personal information and whips them into a custom cake that can be monetized because it describes you to those seeking to reach you. Compared to human editors, software algorithms are scalable, less expensive than hiring humans, and devoid of nuisances such as bathroom breaks, health care, vacations, and sick leave.

Algorithms also unfailingly do what they are told. In the advertising-based digital economy, those instructions are to maximize user engagement so as to display more advertisements, even if that means disregarding its consequences. As Facebook investor and early adviser Roger McNamee explained, "Using a variety of psychological techniques derived from propaganda and the design of gambling systems, Facebook grabs and holds user attention better than any advertising platform before it. Intensive surveillance collects the data that categorize each of its billions of users. Once the data is in hand, algorithms use it to target content that maximizes engagement by pairing topics with user interest and appealing to emotions such as fear and anger."[26]

Outsourcing expense by outsourcing responsibility supercharges the rob-and-rent business model in a way that not only reduces expenses but also increases revenue by releasing the demons of hate and propaganda. Russian efforts to destabilize American democracy, for instance, are profitable for the digital companies because they excite and engage targeted users, who are then shown advertisements. The facts are irrelevant. One set of Russian posts on Facebook or YouTube, for instance, may be anti–Black Lives Matter, while another—from the same source but targeted to a different cohort—may be pro-BLM. The goal of those placing the material is not to inform but to create outrage to threaten social cohesion.

By cutting costs through outsourcing responsibility, social media companies and foreign adversaries have mutually supportive goals: one promotes destabilization while the other profits from it. Nondemocratic, noncapitalistic entities are thus able to exploit unsupervised internet capitalism as a tool to destabilize democratic capitalism.

The capstone of curation for cash rather than truth is how it has destroyed the old belief that the antidote for bad speech is more speech. The alternative voices of "more speech" are silenced by algorithms that curate to deliver only the user's preferences. The effect is the digital tree falling in the forest: if the fact-based, consensus-building message is never heard because the algorithm never delivers it, there is no message.

"COMPETITION IS FOR LOSERS"

"Competition is for losers," legendary Silicon Valley investor Peter Thiel wrote in a classic *Wall Street Journal* article.[27] As an example, he pointed to Google, which, he pointed out, "hasn't competed in search since the early 2000s."

Thiel was an early investor in Facebook. His mentee, Mark Zuckerberg, learned well. In 2008, Zuckerberg explained his strategy for eliminating competition in an email to a colleague: "It is better to buy than compete."[28]

The philosophy was put to work in 2012 when Facebook ended an existential threat by acquiring Instagram, a company that took advantage of Facebook's weakness in mobile devices to develop a popular photo-sharing service. As one journalist commented at the time, Facebook "knew that for the first time in its life it arguably had a competitor that could not only eat its lunch, but also destroy its future prospects."[29]

In 2013 Facebook harnessed data analytics to identify potential competitive threats. For $120 million it acquired two-year-old Onavo and its software to track mobile web activity and identify new services that could pose a competitive threat to Facebook.[30] As *BuzzFeed News* reported, "By acquiring Onavo and turning it into a private tool, the company took away one of the best avenues for understanding mobile trends outside of Facebook's ecosystem. . . . If a potential Facebook killer was on the rise, Facebook could hypothetically spot it before anyone else."[31]

The following year, Onavo data quantified the competitive threat from the mobile messaging platform WhatsApp. "For months, the company had been tracking WhatsApp obsessively using Onavo . . . whose data showed that the messaging app was not just a rising competitor, but a potential Facebook killer," one news outlet reported.[32] In February 2014, Facebook bought WhatsApp for $19 billion, once again demonstrating how "it is better to buy than compete."

The dominant companies have thus established a virtuous circle in which market dominance creates huge stock values and troves of cash that can be used to continue dominance and growth through acquisition. Between 2005 and 2021, Facebook acquired seventy-eight companies.[33] Google chairman Eric Schmidt bragged that in 2011, his company spent $1.4 billion in just twelve months to acquire fifty-seven companies—that is better than one acquisition per week.[34]

The acquisition strategy of the companies has created a new reality for innovators and their investors. The success of innovative new services has become a race between how fast the innovator can reach scale to defend itself by the time the incumbent wakes up and strikes back with a clone or a buyout. It is such a prevalent practice that the venture capitalists who finance the innovators have a name for it: the "kill zone," in which good ideas fall prey to dominant companies.

An example of kill zone activity was Amazon's elimination of a potential competitive threat from Diapers.com. In 2009 a startup named Quidsi began Diapers.com, a site that sold baby supplies. Even though

Amazon did not sell diapers, Diapers.com held the threat of becoming the anchor for an ever-expanding online marketplace—just as Amazon had used books to begin its empire. What happened next was documented by the House Judiciary Committee in its investigation of Big Tech.[35]

By 2010 Quidsi had reached $300 million in revenue. That same year, it started Soap.com to sell health and personal care items. In June 2010 Amazon went on the warpath against Quidsi, suddenly starting to sell diapers at a price that undercut the prices of Diapers.com. This, of course, is a traditional competitive response. It is illegal, however, if such practices involve predatory pricing—that is, a plan to lose money to drive out a competitor so as to raise prices later. Amazon was reportedly willing to bleed cash at a rate of up to an amazing $200 million a month (on diapers!) to take down Quidsi.[36]

As the race for share in the diapers market intensified, Quidsi needed more investment capital to fight the Amazon onslaught. But Quidsi had entered the kill zone, and the venture capitalists were skeptical of putting in more money. In September 2010 the founders of Quidsi flew to Seattle to meet with Amazon to discuss an acquisition. In that meeting Amazon surprised the startup by announcing the creation of Amazon Mom with even deeper price cuts and other benefits. Within two months Quidsi, seeing the writing on the wall, sold itself to Amazon for $545 million.[37]

The diaper story is not unique. As the *Washington Post* reported, "The world's wealthiest companies routinely launch new products free or at money-losing costs that smaller rivals can't manage without going out of business."[38] It is the same strategy John D. Rockefeller used to create the Standard Oil monopoly: price so low that your competitors cannot survive.

But It's Free!

As Rockefeller illustrated in the original Gilded Age, that which is "free" or low cost is not without a price. When we think about Rockefeller and Standard Oil, the common default is to convert "petroleum" to mean "gasoline." Rockefeller's monopoly, however, was built long before the automobile drove demand for gasoline. The staple of Rockefeller's empire was kerosene.

During the Gilded Age, kerosene replaced whale oil in lamps. Not only did it literally "save the whales" but, thanks to Rockefeller, the cost to fill those lamps decreased. A gallon of whale oil cost $1.77 in 1856.[39] The discovery of vast petroleum reserves in the 1860s drove the cost of a gallon of kerosene to twenty-six cents by 1870. At the time, Rockefeller controlled only about 4 percent of the market. By 1890, when he controlled 90 percent of the petroleum business, the cost for a gallon of kerosene had plunged to seven cents.[40]

By reducing the price almost 75 percent, Rockefeller seemed to be almost giving kerosene away. Clearly, there was a consumer benefit as more homes, businesses, and streetscapes were able to conquer the night, and kitchen stoves cooked the family dinner, thanks to low-cost kerosene. This obvious consumer benefit, however, masked Standard Oil's anticompetitive behavior.

A key to Rockefeller's success was his ability to reduce prices to a level that no one without a similar cost structure could match without losing money. The low prices were great for consumers. The market dominance was great for Rockefeller. The result for potential competitors was awful.

Size gave Rockefeller the muscle to reduce costs. At a time when a farmer typically paid forty-five cents to haul a 90-pound can of milk 60 miles by rail, Rockefeller leveraged his role as the largest rail shipper to pay only ten cents to haul a 390-pound barrel of oil over 400 miles and then haul the cars back empty.[41] Even more breathtaking, he was able to coerce the railroads into giving him a rebate on each barrel of oil shipped by his competitors, thus increasing the costs of the competitors while reducing the costs of Standard Oil.

Rockefeller's goal was to drive competitors from the market—and, as Standard's 90 percent market share revealed, it was a successful strategy. One exchange between Rockefeller and an independent refiner sounds like the Amazon-Quidsi "negotiation" of over a century later: "If you refuse to sell [to Rockefeller], it will end in your being crushed."[42]

The Rockefeller experience is relevant to today. Certainly—like today's "free" online services—consumers benefited despite Rockefeller's predatory practices. However, the competitive market was destroyed to such an extent the government ultimately had to step in. In 1911, the U.S. Supreme Court ruled Standard Oil was engaged in a monopolistic restraint of trade, and broke the company into thirty-four independent companies.

That the consumer-facing side of many platform-based businesses is "free" is, as with low-cost kerosene, a wonderful advantage for consumers. But, like the consequences of Rockefeller's activities, it is not costless. The "free" services offered by advertising-based digital companies are catnip to engage consumers so that the companies may siphon away their personal information, package it through algorithms, and then rent the information to advertisers.

THE GRIFT
The secret of the world's greatest business model is its grift. A grifter is a confidence man who promises a reward for something as seemingly simple as tracking a pea under a trio of cups. The grift of the platform companies is a similar come-on that begins with a "free" service.

But what they are "giving away" pales in comparison to what they receive in return: the ability to monopolize the data advertisers value most. Having used "free" to amass an unrivaled store of information, the companies use their dominance in both reach and targeting data to charge what the market will bear and amass even greater profits and political power.

The true cost of "free" can be measured by its effects. For instance, the redirection of advertising dollars away from local media to hypertargeted digital media has the effect of removing support from fact-based journalism in favor of supporting the platforms' non-journalism, often built on non-facts. Local businesses hiring local workers, renting local buildings, and buying local services have succumbed to "free" delivery, meaning their former customers never need to leave home. The price of "free" works today just as low prices did in the original Gilded Age to centralize market power in the hands of a few.

The digital platforms collect, aggregate, and manipulate personal data at marginal costs approaching zero. Then, after hoarding the information, they turn around and charge what the market can bear to those who want to use that data. Their costs to assemble what they monetize are so low as to make the entry of others difficult. But if and when a potential competitor emerges, their profits are so high as to permit the purchase of the potential competitor at an extravagant price. The result is the increasing dominance of the digital platforms.

It is, indeed, the world's greatest business model.

CHAPTER 8

Where Is the Watchdog?

THE GILDED AGE IS GENERALLY CONSIDERED TO HAVE BEGUN WITH
the end of the American Civil War. When the conflict ended in 1865,
it had been twenty-one years since Samuel F. B. Morse's history-making
message "What hath God wrought" and only four years since the Transcontinental Telegraph connected the American coasts. It was the dawn
of the high-speed connectedness that would help drive the era's growth
and would ultimately manifest itself in the internet.

While the telegraph was connecting the vastness of the United
States, in Europe it was leaping national borders. The 1648 Treaty of
Westphalia had defined Europe in terms of lines on a map. Those lines
worked because of geography, hard-to-travel distances, and the ability
to control borders. The telegraph challenged those conditions. Why, for
instance, should messages race across a nation at lightning speed only to
stop at an arbitrary political demarcation?[1]

To deal with the new interconnectedness, in 1865 the French
government assembled representatives of other European nations. The
result was the International Telegraph Union (ITU) and an ongoing
multinational effort for management and oversight of the new communications technology. The ITU was the first supranational organization
in which geography-based nation-states ceded sovereign authority to a
common body.[2]

The supranational ideas behind the ITU started European governments down a path that over time led to a coal union after World War II,
and ultimately the modern European Union (EU). Today, the EU (and

since its departure from the EU, the United Kingdom) are deep into the development of policies to oversee the new transnational communications network of the internet. In the process, they have opened a new contest for power not only among sovereign nations but also with the community of digital technocrats who run the dominant digital companies and in many cases act as their own sovereign.

While the EU and the UK wrestle with the pseudo-governments of the digital giants, China and Russia are trying to impose their will on the internet both within their borders and internationally. Missing from the development of internet policy, however, has been the strong leadership of the United States. Having ceded policy to the digital giants domestically, the vacuum that is internal American policy is being filled internationally.

While the United States is strong in technological development but weak in policy development, in Europe the inverse is true. The nations of the EU see the digital future as essential to their economic future. As a result, they define that future with a set of common rules.

The nations of Europe may share democratic principles with the United States, but they are fierce economic competitors. This has raised the concern that their digital consumer protections may be a mask to hide digital economic protections. Meanwhile, China and Russia, devoid of liberal democratic principles and desirous of economic advantage, are at work to exploit the internet to their own benefit.

WHAT HAPPENED TO AMERICAN LEADERSHIP?

The U.S. government's failure to exercise its sovereignty through behavioral rules for the digital marketplace has, as we have seen, allowed the companies to make their own rules. The ability of the companies to act like their own governments has been aided and abetted by the politics of the new Gilded Age.

Decades of being told "government is not the solution, it is the problem" has fed citizen distrust, activated judicial decisions against governmental intervention, and discouraged congressional action. The result created the optimal environment for the digital companies to frame the issue of government oversight to best advantage themselves.

As we have seen, the Big Con of the consumer-facing digital companies was that the magic of "permissionless innovation" would be destroyed by government oversight. Keeping the government at bay, of course, allowed the companies to assume the policy role for themselves and define their own rules to their own benefit.

To exercise digital sovereignty internationally, however, it is necessary to first demonstrate it domestically. Failure to develop national rules reduces the U.S. role in international policy discussions to debating conceptual perceptions rather than permanent policies.

When the United States developed net neutrality rules in 2015, for instance, it formed the basis for transatlantic cooperation. Two weeks after the Federal Communications Commission (FCC) adopted its net neutrality rules, as chairman of the FCC, I joined FCC general counsel Jon Sallet in London to meet with the twenty-eight EU communications policy regulators (the UK was still a member) as they considered their net neutrality policy.

Together, the representatives of the United States and the EU wrestled with common challenges while respecting national differences. The United State having just emerged from a decades-long debate over such rules, the American experience was highly relevant and a case study from which to work. The ultimate decision was an EU policy that was not identical to but was clearly compatible with the U.S. policy.[3] It demonstrated how it is necessary to bring something to the table in order to enjoy a seat at that table. Unfortunately, after this example of transatlantic cooperation, the Trump administration pulled the rug from under our allies—and American consumers—by repealing net neutrality.

What efforts there have been at American digital oversight have focused on the use of antitrust statutes. The federal government—through the Department of Justice and the Federal Trade Commission (FTC)—as well as state attorneys general have brought lawsuits alleging anticompetitive behavior against the dominant digital companies. Congress in 2022 considered but failed to pass an update to the antitrust laws to deal with digital platform practices. These are important efforts, but there are multiple shortcomings if antitrust is expected to be the principal response to the behaviors of the dominant platforms.

The first such shortcoming is the difficulty of antitrust law to reach important behavioral issues such as the effect of the companies' actions on personal privacy or misinformation. The test, for instance, in the Clayton Antitrust Act, a primary antitrust statute, is whether an activity has the effect to "substantially lessen competition." The other principal antitrust statute, the Sherman Act, is concerned with whether corporate activities are "unfair methods of competition." Broad behavioral issues such as privacy and misinformation thus do not lend themselves to being solved by targeted antitrust enforcement.

Another shortcoming is that antitrust efforts tend to be retrospective in nature and focused on the activities of a single company. Other than reviewing a merger or acquisition, when the government brings an antitrust action, it is an ex post effort based on a harm that is alleged to have already been perpetrated. Because it focuses on a single company's actions, the conclusion is enforced specifically to that company and the identified practice rather than to the broader behavior of all similarly situated digital companies.

Antitrust suits are also reliably lengthy, during which the alleged anticompetitive action continues. The original Justice Department suit against Google, for instance, was filed in October 2020, is scheduled to go to trial in September 2023, and probably will not receive a decision from the U.S. Supreme Court until 2026. In the fast-paced digital world, such a six-year period is an eternity.

The other certainty about antitrust suits is the uncertainty of their outcome. This is especially true when the prevailing jurisprudential precedents have been built around the so-called consumer welfare standard in which the impact on consumers (principally through price) takes precedence over the impact on the market.

All of this is not to suggest that there is not a need for antitrust enforcement and the updating of old competition statutes to reflect the new realities. But society's efforts to deal with the abusive behavior of dominant digital companies must cast a wider net than that afforded by antitrust statutes alone.

The targeted, time-consuming, and uncertain limitations of antitrust suits necessitates agile and ongoing regulatory oversight. Once again, however, the statutes and structure of existing regulatory agencies make comprehensive oversight difficult.

The FCC, for instance, operates on a 1934 statute whose last update was in 1996, at a time when the internet was screeching modems plugged into telephone jacks to access services such as America Online. During the Obama administration the FCC did what it could, imposing net neutrality and network privacy protections on the companies that consumers rely on for their internet connections (both were subsequently repealed in the Trump years). Because the agency's authority was limited to communications networks, however, extending openness and privacy protection requirements to the platforms was not possible.

The FTC arguably has the authority over the platform companies, but its regulatory tools to enact broad-based rules are limited.[4] At the time this book is being written, the Biden FTC, under Chair Lina Khan, is pushing to interpret and implement the FTC's regulatory authority to allow it to deal more broadly with digital consumer protection. This includes a reinterpretation of its rulemaking authority. Chair Khan's leadership is important and hopefully will be decisive. As with any regulatory action, however, the success of such efforts will ultimately be determined by the decisions of judges sitting on appellate courts.

Just as nature abhors a vacuum, so does public policy. The inability of American policymakers to develop a broad set of digital behavioral expectations has invited other nations to provide their own leadership. "A common refrain among European officials," *Politico* reported, "is that they're being forced to take action because the U.S. hasn't."[5]

EUROPEAN LEADERSHIP

The EU and UK are engaging in digital policy jujitsu that combines their own sovereignty with the network's interconnectedness to de facto impose policy decisions on other sovereign states. Leveraging the "we make the rules" arrogance of American digital platforms, the EU and UK have begun their own "no, we make the rules" initiatives.

The case study for the success of European jujitsu is the 2018 imposition of the EU's General Data Protection Regulation (GDPR) to address some of the personal privacy issues created by the digital platforms. The global ubiquity of the internet, where EU citizens link to US-based services and vice versa, has ended up making GDPR the worldwide privacy standard. When, for instance, the state of California enacted its own online privacy statute in 2018, it was based on the GDPR model.[6] Even China has privacy rules that follow the GDPR model (except insofar as the government has access to the data).[7] And while the U.S. Congress has yet to act on digital privacy, whenever it does discuss the topic, GDPR is table stakes.

Of significance to American consumers and companies is that just as in the marketplace, there is a first-mover advantage in regulation. America's inaction has cost it the advantage of such initiative and given it to others.

European Union—Oversight of "Gatekeepers"

The nations of Europe are pursuing collective regulatory solutions through the EU. The effort began with court actions to enforce the competition laws. European competition statutes are considered more strict than American antitrust laws,[8] but the EU's litigation strategy has had mixed results that demonstrate the difficulty of an antitrust-only solution.[9] Even when successful in the imposition of substantial fines, the penalties were gnat bites to the digital giants and failed to result in broad behavioral changes.

In response, the EU strategically repositioned away from after-the-fact enforcement to the establishment of ex ante regulatory expectations.[10] Explained Margrethe Vestager, the EU's executive vice president in charge of digital and competition policy, "We also need new ways to tackle the problems that digitization causes."[11] That new approach has been to focus on the development of broadly applicable behavioral rules. One set of rules focuses on market-related issues, another addresses content-related practices, while a third addresses artificial intelligence.

The EU's Digital Markets Act (DMA)[12] looks at the marketplace behavior of companies that are determined to be gatekeepers. These are enterprises that, because of their systemic importance and strong economic position, possess an entrenched and durable posture allowing them to control the market and the activities of other companies. A company is determined to be a gatekeeper based on three criteria: (1) size (annual revenue or market value), (2) user base in the EU, and (3) the conclusion that triggering the two preceding criteria will create an "entrenched and durable position" in the market.[13] The vast majority of companies fitting the gatekeeper definition are located outside the EU, predominantly in the United States. For these gatekeepers, the EU imposes an array of behavioral mandates, including transparency, third-party interoperability, data portability, opt-in consumer authority, the ability for consumers to uninstall bundled applications, and the prohibition of self-preferencing on their own platform.

When it comes to content delivered by the internet, the EU's Digital Services Act (DSA)[14] bans content deemed illegal or harmful under European laws. Such a determination reflects how the protection of free expression is more limited in Europe than in the United States.[15] For instance, the DSA requires companies to document how they will separate illegal content—such as hate speech—from free expression and stop it.

The Artifical Intelligence Act (AIA), as of this writing, has been adopted by the European Council (representing nation-states) and the European Parliament (an elected body). The details of its implementation are being resolved between those two bodies and the administrative European Commission. Two things stand out in regard to the EU's AI oversight: the formal processes to develop the policy began years ago (unlike in the United States) and the oversight is based on risk analysis rather than on industrial-style micromanagement.

Whether it is market behavior, online content, or AI, the European Commission is beginning its regulatory implementation. While the specific policies are important, the broader significance of the EU's activities is how the investment of time and resources in the issue of digital oversight outstrips anything done in the United States and reflects the willingness of other liberal democracies to insert themselves into the operation of the digital marketplace.

United Kingdom—Oversight of "Strategic Market Share"

Having withdrawn from the EU, the UK is charting its own course for digital oversight.

The UK's conclusions about the marketplace effects of digital companies are similar to the conclusions of the EU, including that antitrust, while useful, is not sufficient. Yet the desired regulatory oversight proposed in the UK is different from that on the other side of the English Channel.

The UK's policy development began with the March 2019 government report, *Unlocking Digital Competition: Report of the Digital Competition Expert Panel*, chaired by Jason Furman.[16] It recommended focused attention on specific companies, especially those driven by advertising and other transactional revenue, in a manner specific to their activities. While the EU focuses its regulation on gatekeepers, the UK proposal focuses on companies with "substantial market power." The UK policymakers concluded that an EU-like determination of gatekeepers is a "mechanical" determination (based on size, revenue, etc.) that ignores divergent corporate behaviors rooted in unique applications of substantial market power.

The UK approach empowers an existing agency of government—the Competition & Markets Authority (CMA)—to put the initiative in place. The CMA, in turn, created a new Digital Markets Unit (DMU) to develop and enforce bespoke behavioral expectations for each target company and to review the merger and acquisition activities of the companies.

Like the EU, the UK has separated content-related issues from marketplace-related issues. Under its Online Safety Bill, the broadcast regulator Ofcom (Office of Communications) has received new authority to establish court-enforceable codes of conduct for information providers.

The UK is also exploring the oversight of AI, including the establishment of the UK as the hub for international regularory cooperation on the topic.

A Matter of Priorities

Regardless of the nuanced differences between EU and UK policies, both prioritize ex ante efforts to prevent harms. The United States, on the other hand, appears to be awaiting an accumulation of harm before acting.

At the core of the transatlantic differences are disparities in prioritization. In the United States, policy tends to prioritize the economics of the companies as national champions. The Europeans, with few dominant companies within their borders, prioritize the marketplace effects of the companies' actions on consumers and competition.

The challenge in an interconnected world, however, is how the de jure policies in one nation (or collection of nations) can become de facto reality where a government has failed to create its own standards.

China's New Rules

At a time when the United States is struggling to determine the proper oversight of digital platform companies, the Chinese government has moved decisively forward with its own plan. The authoritarian manner in which the new policies were enacted is antithetical to Western values. Yet the Chinese policies are strikingly similar to policies being discussed on both sides of the Atlantic. While the United States talks about digital policies but does not act, the EU moves to implement transnational regulations, and the UK looks to the authorities of existing agencies, China has acted.

U.S. policymakers continue to debate the acquisition practices of tech firms. China has blocked mergers.[17]

In the United States, sanctions against digital companies face a lengthy uphill battle in court. China simply imposed a record fine on Alibaba for its alleged abusive actions.[18]

Western digital platforms continue to operate "walled gardens" to keep potential competitors from interconnecting with them. China, while protecting its companies behind the Great Firewall, ordered its dominant platforms to interconnect with rivals.[19]

Open access to data and the user's right to data portability are discussed in the liberal democracies. China has ordered its companies to share their data and allow consumers to take their data with them if they change services.[20]

American digital platforms argue they are an essential contributor to national security through their tech development activities. China's actions to regulate their platform companies challenge that assertion by distinguishing between the innovative capabilities of "consumer tech" and defense-related "deep tech" companies.[21]

The decision-making process of liberal democracies is certainly more open, deliberative, and time-consuming than in China. Yet the fact that China is first across the finish line with digital platform policies establishes its own first-mover advantage. Just as the EU's GDPR became a de facto international standard (even forming the basis of China's new privacy rules), the actions of the Chinese government, coupled with the interconnectedness of the internet, will add Chinese policies to the mix of international regulatory standards.

WATCHDOGS PROWL

The international interconnectedness of the internet has allowed nation-states to extend their sovereignty beyond their borders. At the same time, the network has empowered the dominant platform companies to assert a new kind of technological sovereignty that subsumes borders.

The conflict is coming to a head. Platform technocrats are behaving as though the network made them more powerful than nations. Supranational unions and nation-states such as the EU, UK, and China, however, are pushing back and see the network as providing a transnational multiplier effect to extend their sovereignty beyond their borders to dictate behavior in other markets.

These international actions will affect American companies. In its 2021 annual report to the U.S. Securities and Exchange Commission (SEC), Meta Platforms, Inc., warned, "If we are unable to transfer data between and among countries and regions in which we operate, or if we are restricted from sharing data among our products and services, it could affect our ability to provide our services." If policy agreements cannot be reached, the filing stated, "we will likely be unable to offer a number of our most significant products and services, including Facebook and Instagram, in Europe."[22] While the language is a boilerplate response to the SEC's requirement to explain potential adverse developments, it nonetheless highlights the face-off between the watch-dogs and the platforms.

Watchdogs, by their nature, prowl. The internet has allowed the plat-forms into new neighborhoods. At the same time, the watchdogs of those neighborhoods have been given the opportunity to similarly utilize the internet to expand their territory.

PART IV

REASSERTING THE PUBLIC INTEREST

"We didn't take a broad enough view of our responsibility."
MARK ZUCKERBERG, BEFORE U.S. SENATE 2018[1]

CHAPTER 9

Designing Behavioral Expectations

THE TIME HAS COME FOR AMERICAN POLICYMAKERS TO REASSERT THE public interest over the interest of the digital platforms. To accomplish this requires policymakers to move from industrial-era thinking to the kind of forward-looking innovation embraced by the digital entrepreneurs.

Thus far, the policy debate has centered on whether platform companies are making the correct management decisions, especially in regard to privacy, competition, and truth. The companies, for the most part, have responded with "we'll do better." It is, however, the wrong question and an insufficient answer.

The question to be answered sooner rather than later is whether such matters of great consequence should be the exclusive decision of the companies in the first place. In the alternative, there should be guardrails to protect the public interest by establishing behavioral expectations for the activities of digital platform companies.

Twitter founder Jack Dorsey's observation that Silicon Valley would have probably built a much better internet and social media world if it had involved social scientists alongside computer scientists in the development of its products is right on target.[1] The evolution of digital technology into business plans too often focused on "Hey, let's see if we can . . ." absent a consideration of what the consequences of those actions might be.

The engineering decisions of digital entrepreneurs have created a series of harms to consumers and competition. Just as those innovators pioneered new ways of thinking about the application of technology, so must policymakers pioneer new ways of protecting against the adverse effects of those actions.

When clever engineers and striving entrepreneurs make the rules, the result is one-sided: decisions that focus more on what the technology can do to drive revenue than on the effects of those decisions on individuals. Mark Zuckerberg put his finger on the challenge when he told Congress in 2018, "We didn't take a broad enough view of our responsibility."[2]

It is not that the platform companies are unable to add behavioral sensitivity to the design of their products and business models. Such consideration, however, is often dwarfed by fascination with the capabilities of digital technology and the lure of monetary returns. Pursuit of these incentives absent an equivalent consideration of the public interest has been the norm because of the failure of policymakers to insist the companies consider the impact of their decisions and provide a suitable structure for achieving their broader responsibilities.

The platform companies seized on their first-mover advantage to design the technical aspects of their products. The challenge now is to create a counterbalancing incentive that will help companies include in their business practices what Zuckerberg called "a broad enough view" of their responsibility.

The following three chapters explore long unaddressed issues that only become exacerbated by the introduction of AI and the metaverse. Ideas to institute privacy protections, promote competition, and protect truth on the dominant digital platforms have yet to be meaningfully instituted for the services we take for granted today. The introduction of next-generation services such as the metaverse and AI will not only intensify the unresolved issues but also complicate their resolution.

The current efforts to "fix" the adverse consequences of the innovators' decisions with an amendment here or a regulation there are laudable but inadequate for the long term. Such actions, as well intentioned as they are, deal with the symptoms of the problem, not the contagion itself. What is needed is the institutionalization of a new ethos that adds a public interest component to how products are designed and implemented.

Like digital technology itself, this oversight will need to evolve with technology and the times. The companies release products that continue to be improved based on new technology and market developments. Oversight of those actions will need to be similarly agile.

Because the consequences of the activities of the digital platform companies are varied and multiple (affecting, e.g., privacy, competition, and the accuracy of information), there cannot be a single meta-solution (e.g., antitrust enforcement). But there is a common concept that can be applied to all the behavioral effects: the expectation that the mitigation of ill effects will be a forethought rather than an afterthought.

In "New Digital Realities: New Oversight Solutions in the U.S.,"[3] Phil Verveer, Gene Kimmelman, and I proposed an approach to addressing the public interest challenges of the digital era. We proposed the creation of a new federal expert agency, the Digital Platform Agency, with a new approach to public interest oversight. While the headline was a new agency, at the core of the proposal was a new regulatory paradigm.

Decades of personal regulatory experience led us to conclude that the current sclerotic and rigid model of regulatory oversight does not meet the needs of the digital era. In its place, we proposed a new approach based upon the process the tech companies utilize to develop *technical* standards. In this case, however, that process is applied to the creation of enforceable *behavioral* standards in the design of digital platform services.

When platform companies are the consumers of tech products, they work together to design a standard for those products that anticipates and mitigates unintended operational consequences. Such a "by design" effort to identify and mitigate potential adverse effects in the consumer marketplace, however, is not a part of this discipline. To instill such an expectation would help rebalance the company-consumer relationship.

A requirement to apply such a "by design" concept to the oversight of the digital consumer market does not mean "by dictate" as did industrial era regulation. Rather than micromanagement, the new model should embrace a supervised, cooperative process of risk management to develop enforceable codes of conduct to mitigate those risks.

Throughout the following three chapters the theme is to move beyond pruning the effects of digital technology in favor of getting to the root of the issue: how digital products and services are designed and implemented by the dominant digital providers. It is time to design for privacy . . . design for competition . . . and design for truth.

NEGATIVE EXTERNALITIES

Industrial polluters and the dominant platform companies share a common characteristic: their behavior increases corporate profits by transferring costs to others. The early twentieth-century British economist Arthur Pigou[4] described such practices as producing "externalities"—the effects generated when a business does not absorb all the costs associated with its activities.[5] Classic examples of negative externalities are the carcinogenic effects of environmental pollution and tobacco products in which the cleanup and/or health effect costs are passed to individuals and society.

The digital platforms have created their own class of negative externalities. The capture of personal private information, often unbeknownst to the individual, transfers privacy data protection costs to the individual. The decision of digital platforms to curate their content for maximum engagement results in negative externalities ranging from bullying to lies, hate, and disinformation campaigns by foreign governments. The chapters on the metaverse (chapter 4) and AI (chapter 5) outline the new negative externalities those activities bring with them.

Pigou argued that the existence of such negative externalities was prima facie justification for government intervention.[6] In particular, he favored using tax policy to either encourage the company to resolve the problem itself or to fund society's mitigation efforts.[7] What is needed in the case of the dominant digital platforms, however, reaches beyond the imposition of taxes. The decisions of the platforms to maximize for corporate return by imposing costs on their users and society as a whole demands reprioritization through the establishment of behavioral expectations for those companies.

RESUSCITATING COMMON LAW CONCEPTS

A basis for establishing behavioral expectations for digital platforms already exists. It is the same underpinning that in the Gilded Age helped rebalance the marketplace between commercial might and consumer rights.

In developing policies for the industrial age, legislators and judges looked for guidance to the concepts of English common law. Those same time-tested standards can also form the basis for oversight of the digital platforms.

As England was breaking free of the bonds of feudalism, a set of principles arose to protect the nascent middle class. It was referred to as "common law" because, as opposed to the random rules nobles had imposed on serfs for centuries, these principles were "common" in courts throughout the land.

At the core of common law is precedent that reaches back to the twelfth century to define acceptable behavior. Imported into the American colonies (and others touched by the British Empire), common law continues to form an important part of American jurisprudence.

Common law is also looked to for guidance by legislatures. When Senator John Sherman advocated for the antitrust statute that today bears his name, for instance, his key argument was how it was the simple extension of common law to the new industrial economy.[8]

There are at least two time-tested common law principles applicable to the design of behavioral expectations for the digital age.

One applicable concept is the Duty of Care. This holds that the provider of a good or service has the obligation to anticipate and mitigate potential harms that may result.

Among its other applications in the industrial age, the Duty of Care established behavioral expectations for the new network technology of the railroad. As nineteenth-century trains raced across farmers' lands, the steam engines threw off hot cinders that would set fire to the barns, hayricks, and homes as they passed. The Duty of Care, in the form of the tort claim of negligence, was enforced against the offending railroads. The result was that the railroads installed screens across the smokestacks of the steam engines to catch the cinders.[9]

The digital economy needs digital smokestack screens to catch the dangerous effects thrown off by platform companies.

The other common law concept applicable to today is the Duty to Deal. Under this holding, the providers of a critical service have the responsibility to make that service available to all on a nondiscriminatory basis.

When Congress, in 1887, created the first independent regulatory agency, the Interstate Commerce Commission (ICC), to oversee the railroads, it implicitly embraced a nondiscriminatory Duty to Deal. The Interstate Commerce Act of 1887 provided "That it shall be unlawful . . . to make or give any undue preference or advantage to any particular person . . . or subject any particular person . . . to any undue or unreasonable prejudice or disadvantage in any respect whatsoever."[10]

This nondiscrimination principle was explicitly extended to common carriers such as telephone networks in the Communications Act of 1934: "It shall be unlawful for any common carrier to make any unjust or unreasonable discrimination."[11] Extending nondiscrimination to the delivery of the internet was also at the heart of the Obama FCC's net neutrality rules.[12]

Unfortunately, the timeless principles of common law have been hit by the blitzkrieg of the digital revolution. As digital innovators made their own rules, they often conveniently sidestepped or blatantly ignored the duties to others embedded in common law.

The behavioral principles for the digital era already exist. The challenge is how to transfer these common law responsibilities into twenty-first-century policy.

THINKING ANEW

Former secretary of state Madeline Albright once made an observation about modern diplomacy that can be equally applicable to technology policy. Twenty-first-century problems, she explained, were too often discussed in twentieth-century terms, and dealt with using nineteenth-century solutions.[13]

Technology policy is stuck in the same conundrum. Policymakers quite naturally have tended to interpret digital developments based on their understanding of the world in which they matured. The twenty-first century digital economy, however, requires a new perspective that produces new policies.

Thinking about how industrial age statutes and regulatory structures can be used to trim back some of the activities of the dominant digital platforms is not sufficient. It is those statutes that have allowed the digital platforms to make the rules. It is time to reassert the sovereignty of We the People and approach new challenges from a new perspective.

As we have seen, industrial activity was a linear process in which hard assets were fed into a step-by-step pipeline to produce tangible products. In contrast, digital platforms are all about the pairing of soft digital assets to produce intangible results with exponential returns.[14] The policies that worked to apply basic fairness and competitive concepts to the production economy do not always successfully extend to oversight of the pairing economy.

What is required today is a reprise of the thinking that rebalanced the marketplace from its Gilded Age excesses. The policy initiatives of the Gilded Age had to collectively extend beyond the policies developed for the prior agrarian economy. The policies of the digital era must exhibit similar creativity to extend beyond the policies developed for industrial activity.

Theodore Roosevelt explained it this way to Congress toward the end of the Gilded Age: "The old laws, and the old customs which had almost the binding force of law, were once quite sufficient . . . [but] since the industrial changes which have so enormously increased the productive power of mankind, they are no longer sufficient."[15] Substitute "digital" for "industrial" and the observation breathes anew.

The digital changes that have "so enormously increased the productive power of mankind" have made the old policies "no longer sufficient." The development of policies sufficient for today's needs requires looking beyond what has worked before.

YOU LOOK LIKE YOUR PET

The regulatory policies of the Gilded Age were developed in response to the activities of the era's major industrial companies. To manage the process, the government adopted the management concepts of the companies themselves. It was a classic situation of "you look like your pet."

The networks and production platforms of the nineteenth and twentieth centuries centralized economic activity. The world of mass production, mass consumption, and mass media was built by the industrial efficiencies of mass. Factories combined masses of material with masses of workers to produce masses of identical products for delivery to a waiting mass market. The media were all about mass efficiencies as well, with a limited number of major outlets feeding a mass of readers, then listeners, then viewers.

The oversight of such at-scale activities introduced a new skill into the economy: management at scale. As discussed in chapter 2, the management techniques that had previously sufficed for small enterprises were no longer appropriate. A small shop of artisans was a self-managing proposition. A factory with hundreds of workers, let alone multiple factories with activities coordinated by telegraph, required a new level of management.

The management guru of the industrial era was Frederick W. Taylor. The term "Taylorism" became a common description of the prevailing management technique. In *The Principles of Scientific Management*, Taylor wrote, "It is only through *enforced* standardization of methods, *enforced* adoption of the best implements and working conditions, and *enforced* cooperation that . . . faster work can be assured."[16] That Taylor italicized the repeated use of the word "enforced" tells us what we need to know about the management practices of the era.

When the legislators of the late nineteenth and early twentieth centuries acted to curb corporate abuses, they copied the corporate world's command-and-control model. Specifically, these rules mirrored two corporate management methods.

The first was top-down rules. A worker's job on the factory floor was to follow specific rules handed down from on high that dictated the performance of a specific task. The oversight of industrial output was similarly based on top-down mandates.

The second practice imported from the corporate world was bureaucracy. To check that the worker was indeed following the rules was the responsibility of a supervisor. That supervisor was in turn overseen by a manager and corporate hierarchy. Government regulation also came to rely on bureaucracy.

Such a hierarchical, rules-based structure no longer prevails as the management technique of the digital age. Corporate management today has been forced to become more agile and responsive to the rapid pace of technological and marketplace developments. For government to cling to management techniques of the Gilded Age that are no longer relevant in the digital age is harmful to both corporations and consumers.

AGILE OVERSIGHT

Industrial era regulation of networks and services was often built around so-called utility regulation that precisely dictated corporate behavior, including prior approval of prices, investments, and other decisions. Such micromanagement was made possible by the slow pace of technological change and slow marketplace assimilation of that technology.

The fast pace of the digital era means that such command-and-control regulation can be counterproductive. The goal of regulation should be to protect the public interest while encouraging boundary-expanding innovation. Yet, while the pace of the digital era has forced corporations to evolve their management style to embrace agility and continuous development, government management has not kept up with that change.

A new digital product is introduced, for instance, as an "MVP"—"minimum viable product"—with the expectation that it will be continually improved as technology and the marketplace evolve. Every time our smartphones receive a new software update, it is a manifestation of agile management and MVP at work. Government oversight, however, is notably unagile.

In the corporate world, Schumpeter's law of creative destruction[17] drives an environment in which the status quo is fraught with danger. In the government world, where there is no such incentive, status quo business as usual is too often the safer path. Taking risk in the for-profit sector is expected. Taking risk in government is often rewarded with investigations, headlines, congressional hearings, and pushback from politically powerful affected parties.

Regulation no longer "looks like your pet" because the pet has evolved in response to agile management methods while the regulators have remained stuck in rigid, top-down practices developed in the slower-paced industrial era. Oversight of digital activities calls for replacing the dictation of specific operational details with dynamic regulation that is focused on identifying and mitigating significant risks with agile solutions.

The industrial regulatory model may no longer serve the interests of the public, but it does serve the interest of those opposed to regulation. A go-to argument of the antiregulation forces is that the rigidity of regulation inhibits the spontaneity of innovation. Corporate management moved from rigid rules to agile responsiveness precisely because the old system slowed innovation at a time when technological change meant innovate or perish. When government does not similarly evolve to embrace agile regulatory concepts, it creates a credible anti-innovation argument for those opposed to regulation in any form.

Accomplishing the necessary agility within a federal agency is difficult. My experience as chairman of the FCC taught me how hard it is to implement agile management practices within industrial era statutes.[18] The process created to ensure the full and complete representation of views—an essential activity—was lengthy and more suited to those with large legal teams than to the public. Decision-making based on the information so gathered has become increasingly politicized. And, when those unhappy with the decision file suit to challenge it—as is their right—the ultimate decision ends up in the hands of judges of general expertise— another reliably lengthy and rigid process.

What is needed in the digital era is not the absence of oversight but regulation that, like the entities it supervises, is focused, agile, and responsive.

Government oversight must once again "look like your pet." This means, where possible, adopting agile techniques that have proven successful for the digital companies.

CREATING DIGITAL CODES

One of the tools digital companies use to keep up with the rapid pace of technological advancement is ever-evolving technical standards. Through formal industry-organized groups, technology rights holders, device manufacturers, and service providers collectively develop an endless stream of technology standards, resulting in ever-evolving products and services.

The evolution, for example, of mobile communications standards—from first generation (1G), to 2G, to 3G, to 4G, to 5G, and the ongoing development of 6G—is an agile process that continually redesigns products to reflect the capabilities of new technology and the needs of the new market.

The process by which technical standards keep pace with change provides a model for a similar establishment of agile and responsive behavioral standards.

There is a bountiful supply of private organizations that bring companies together to agree on technical standards about how things will work. The jack that plugs your phone into the wall is the result of standardization. You can take your mobile phone anywhere in the world because the networks are standardized. And, of course, the internet itself is the result of standards that allow disparate networks and devices to work together.

Absent from such technical standard setting in the digital world, however, has been behavioral standards for what the technology enables. The law of property, contract, and tort provides some standardized legal expectations, but these are not sufficient, as we have seen, to contend with the issues brought forward by digital technology.

Self-regulation is a step in the right direction, but it is insufficient. I once ran the process to develop a voluntary self-regulation code for the wireless industry. It was the responsible thing for the industry to do. There are two problems with such voluntary efforts, however. The first limitation is that the code is only as strong as the weakest link in the group developing it. The price of consensus is often the absence of an optimal conclusion. The other shortcoming is the lack of meaningful enforcement. That which is voluntary is per se unenforceable.

The traditional approach of government has been to wait until abuses are apparent and then to apply concepts that worked previously.

We need not wait any longer; the abuses are abundant and apparent.

The regulatory concepts that worked in the industrial era cannot keep pace with, or embrace the innovation of, the digital era.

A government-overseen process to establish agile behavioral codes based on common law principles would be a realistic new approach to tackle the consequences-free behavior of the dominant digital companies.[19]

FROM MICROMANAGEMENT TO RISK MANAGEMENT

To reapply "you look like your pet" parallelism between digital management and oversight requires retooling the role of federal regulators in a manner that reflects the dynamism of digital management. This begins with substituting risk management for micromanagement. So-called utility regulation with its detailed direction of corporate operations does not work in the digital era of fast-changing technology and markets. Digital oversight requires a switch from overseeing "how" a company operates to "what" the consequences of those decisions may be.

Such an approach can be best achieved through the establishment of enforceable industry codes utilizing the kind of multistakeholder process developed for the creation of technical standards. That the platform companies have been unable to accomplish this themselves demonstrates the need for government intervention. The new process requires that government redefine its purposes and procedures.

Such a recasting of the role of government from a dictator to an orchestrator with enforcement responsibilities has been successful in other segments of the economy. The Financial Industry Regulatory Authority (FINRA) regulates aspects of the financial markets through an industry-developed code overseen by the Securities and Exchange Commission (SEC). The North American Energy Reliability Corporation (NERC) is an industry-led group that has developed policies to prevent blackouts and is overseen by the Federal Energy Regulatory Commission (FERC).

The development of behavioral codes for digital platforms would include:

- *Orientation toward Results*—The process must begin with the expectation that a meaningful behavioral design will result. The responsible government agency should identify the issues(s) to be addressed and establish a timeline for the results. To keep from falling into the weakest link trap, the final decision should be ratified and/or amended by the responsible agency.

- *Risk Identification*—The responsible agency should begin the process with its own detailed report on the problematic behaviors, along with potential remedies. This analysis should be the "prosecution's brief" and should identify and quantify the issues to be addressed. The focus of the specific code development process then becomes producing an outcome that solves or mitigates the identified problem(s).

- *Multistakeholder Involvement*—Those participating in the process must represent a cross-section of all the interested or affected parties. The participants must have relevant expertise and decision-making authority. The group should be selected by the responsible agency to represent both industry and civil society. Representatives of the agency should be included in the proceedings, with full participatory rights.

- *Meaningful Enforcement*—The appropriate federal agency must have the authority on its own initiative to enforce the standard. After the agency approves or alters the multistakeholder group's decision, it becomes an enforceable policy.

SPECIFIC DESIGNS FOR SPECIFIC CHALLENGES
To give digital innovators the benefit of the doubt, they were developing into the unknown. Even with this dispensation, however, the dominant digital platforms have failed in their responsibility to adequately anticipate and mitigate the broad societal consequences of their actions. The

U.S. government cannot commit the same infraction by failing to consider the consequences of its (in)action.

Common law principles provide a time-tested framework for the establishment of digital design responsibilities. Implementing that framework in a manner that once again "looks like your pet" learns from the management behavior that encouraged digital innovation in the first place.

Responding to the behavior of the dominant platform companies requires the establishment of a countervailing force, not a dictatorial force. Such oversight, if it is to keep pace with the accelerated impact of AI, will also need to embrace AI both to moniter marketplace activity and to assist in regulatory development and enforcement.

The focus of such an activity should be the dominant digital companies—those that are systemically important to their ecosystem. It is the decisions of these big companies that de facto establish the behavior of the smaller companies. Focusing on the systemically important companies means prioritizing those whose practices set the rules, not the insurgent startups. Responsible practices must begin with those who possess marketplace power, and the rest will follow.

The digital revolution has delivered wondrous new products and services while overwhelming the old processes for dealing with an imbalance in marketplace power. The goal of the new process should be a consultative and enforceable code based on the historically established values of common law applied to the realities of ever-evolving technology and ever-changing markets.

The following three chapters discuss concepts for "behavior by design" built around such a new oversight model and how it could deal with some of the most pressing consequences imposed on society by the dominant digital platforms. Dealing with such consequences—the invasion of privacy, the lack of marketplace competition, and the flood of misinformation—only becomes more imperative as we stare down the barrel of AI.

CHAPTER 10

Privacy by Design

IT IS A FEDERAL OFFENSE TO OPEN ANOTHER PERSON'S MAIL TO EXTRACT personal information.[1] It is legal, however, for the providers of internet services to extract personal information when users go online.

Policymakers' inaction has resulted in private personal information that is protected in its analog form becoming unprotected when it is digital. This policy stasis has once again allowed the digital businesses to make their own rules. The result has been previously described as a kind of digital alchemy where private information is converted into a corporate asset that is used to make decisions that affect people's lives.

No wonder 64 percent of Americans think the government should do more to regulate how internet companies handle privacy issues![2] It is not surprising that 70 percent lack the confidence that their personal data are private and safe from distribution without their knowledge.[3] When a whopping 95 percent of Americans believe their personal privacy "now rivals our Constitution's basic rights," online privacy concerns have reached escape velocity.[4]

In the Gilded Age the extraction of industrial assets from the earth created physical wastelands as hills were leveled and pits dug into verdant landscapes. The extraction of the twenty-first-century's capital asset also brings with it an environmental cost—this time, however, it is the human environment. Mining for minerals has been replaced by data mining at an exponential scale. The extractive resource is now people and the private personal information that distinguishes each individual.

Multiple policy solutions have been advanced, yet they typically focus on remediation of specific problems. Ultimately, it is necessary to go to the root of the problem to attack its negative externalities rather than simply trimming some of its negative ill effects. This means providing engineers and entrepreneurs with standards that establish that privacy must be a design component of digital products and services. The time is now to get in front of the privacy challenges—both present and future—to replace "we make the rules" with the expectation that products and services will be designed to protect the privacy of users.

THE SHORT LIFE OF AMERICAN PRIVACY RULES

For years it has been against the law for a telephone company to disclose whom a caller has dialed (absent a court order).[5] In 2016, the FCC extended that concept to the networks that deliver the internet. Two months after the Trump administration took over in 2017, the Republican-led Congress passed a law repealing those privacy protections.

Since the networks that carry consumers to and from the platforms can see everything a user is doing, they are in a powerful position to collect information about individuals' web browsing, the apps they are using, their location, and other personal information. The FCC rule required the networks to get permission from their customers before tracking their online activities. The rule also required the companies to use "reasonable measures" to protect consumers' data from breaches.[6]

"There is a basic truth: It is the consumer's information," I explained when the FCC adopted the privacy rule. "It is not the information of the network the consumer hires to deliver that information."[7]

The companies hated it—even the platforms that were not covered because of limits on the FCC's jurisdiction. Collectively, the networks and platforms launched a campaign that resulted in the congressional repeal which President Trump signed into law.

It was a campaign based on misinformation. "It is unnecessary, confusing and adds another innovation-stifling regulation," Senator Flake (R-AZ), sponsor of the repeal, argued on the senate floor.[8] It was not a new argument—nor was it factual. As the FCC was considering the privacy protections, Senator Flake called a hearing to explore what we were doing and express his opposition. At that hearing, he raised the same "confusion" argument. I replied it was just the opposite. Taking my smartphone from my pocket, I explained how if I were to use it to make a phone call to Air France, the FCC rules would prohibit the phone company from selling that information to a tour promoter in Paris. But if I used the same device operating on the same network to go online to the Air France website, that information was not protected. *That*, I tried to explain, was the confusion that was being avoided and the privacy being protected.

Because it is politically unacceptable to argue "the companies don't like this," the old Washington adage that "you can't kill something with nothing" went to work. The repeal was misrepresented as a pro-consumer effort. Suddenly, repealing privacy protections was actually protecting consumers!

Over on the House of Representatives side of the Capitol, Representative Nancy Pelosi (D-CA) focused on how repeal would hurt consumers. She did this using a powerfully succinct chart headlined "Republicans want this information to be sold without your permission."

COURTESY OF C-SPAN.

The chart listed: "The websites you visit; The apps you use; Your search history; The content of your emails; Your health & financial data"—all would be available for the networks to collect and monetize as they watched everything a user was doing online.

The Republican-controlled Senate passed the repeal on a party-line vote 50–48.[9] The House passed the privacy repeal 215–205 (with 15 Republicans joining the Democrat opposition).[10] The result left the United States without any meaningful rules to protect the privacy of internet users.

Once a champion of individual rights such as privacy, the United States has lost its international policy leadership role to Europe and even China. The European Union's 2018 General Data Protection Regulation (GDPR)[11] has become the de facto international standard.

Because of the interconnected nature of the internet, American consumers are receiving secondary privacy protections from the European rules if the service they use also has European customers. The privacy rights of Americans, however, deserve primary, not secondary, status and protections.

WHEN GOOGLE DIDN'T LIKE ADVERTISING

The heart of the digital platform business model is not the creation of products and services, but the siphoning of private personal information that can be rented to advertisers. It is worth repeating the anecdote from this book's preface about Google co-founder Larry Page's response to the question, "What is Google?" His response: "If we did have a category, it would be personal information. . . . Sensors are really cheap. . . . Storage is really cheap. . . . Everything you've ever heard or seen or experienced will become searchable. *Your whole life will be searchable* [emphasis added]."[12]

The most lucrative—and fastest—way to monetize the information online platforms collect about us is through selling access to "your whole life." Advertisers and others pay handsomely to use this data to micro-target their messages. Interestingly, in the early days of Google, its founders worried about the negative impact of advertising on their search engine and consumers.

"In this paper, we present Google," Sergey Brin and Larry Page wrote in their 1998 "Anatomy of a Large-Scale Hypertextual Web Search Engine."[13] After describing how Google works, the paper concluded, "Currently, the predominant business model for commercial search engines is advertising. The goals of the advertising business model do not always correspond to providing quality search to users . . . we expect that advertising funded search engines will be inherently biased towards the advertisers and away from the needs of the consumers."

The following year, venture capital firms Sequoia Capital and Kleiner Perkins invested $25 million in Google.[14] The year after that (2000) Google unveiled its AdWords platform in which advertisers bid to place their messages in Google searches.[15]

The year before AdWords, Google's revenues were $220,000. The year after AdWords, revenue jumped to $86 million.[16] In 2007 the company bought DoubleClick, a pioneer in the use of tracking cookies, for $3.1 billion.[17]

It was Google that invented "surveillance capitalism," Shoshana Zuboff, who coined the term, has written.[18] "Instead of selling search to users," Zuboff explained, "Google survived by turning its search engine into a sophisticated surveillance medium for seizing human data."[19]

The economics of siphoning personal information to be repackaged as a product for advertisers are simply too powerful for a company like Google (now renamed Alphabet, with a plethora of information-collecting subsidiaries) to ignore. The search engine called Google tracks everything from your web browsing to your physical location.[20] The video platform YouTube, purchased by Alphabet/Google for $1.65 billion in 2006, is integrated with those elaborate corporate tracking capabilities.[21] The Alphabet-owned Android mobile operating system can track what goes on in almost 80 percent of the world's smartphones.[22]

It was a slippery slope from when the use of personal information to target advertising was considered to bias activity "away from the needs of the consumers" to "your whole life will be searchable." But it sure can be profitable. At the start of 2022, Alphabet was the fourth most valuable company in the world, with a market capitalization approaching $2 trillion.[23] The success of the personal information-based business model has not been lost on others. At the beginning of 2022, five of the seven most valuable companies in the world—Alphabet, Apple, Microsoft, Amazon, and Meta (Facebook)—all profit from the collection and use of personal information.

PERSONAL INFORMATION: THE ENTRY DRUG

Beyond the use of personal information to generate profits is the effect that great hoards of personal data have on marketplace competition and the distribution of lies and hate (topics discussed in the following two chapters). Having aggregated vast amounts of personal information, the platforms turn it into a weapon against competitors and a controlling force in the distribution of news and information.

Because the data collected about individuals is the currency of the platform economy, whoever has the most data can affect market competition by denying it to others. Almost three-quarters of all U.S. advertising in 2022 went to digital media.[24] Of that, half is controlled by Google and Facebook.[25] Such dominance is the result of the micro-targeting permitted by the vast collection of information these companies hold on their users. Denying others access to that data, the companies are able to deny potential competitors the tools necessary to successfully compete. If an advertiser has a choice between the greater targeting precision of the big data set or the less accurate precision of the competitor, the decision is obvious.

By denying others access to data, the companies create a bottleneck that maximizes the value of their data holdings. As has been discussed previously, the original Gilded Age saw "monopoly rents"—the additional amount a company can charge because there is no alternative. In the new Gilded Age, the same scarcity premium becomes "algorithmic rents"—the ability to charge revenue-maximizing prices because no one else has the data to offer such algorithmic precision.

The control of private data is also felt in the quality of information delivered to users. Social media platforms apply the same kind of targeting they sell to advertisers to the information distributed in their newsfeeds. This time it is for the purpose of controlling the user's behavior to keep them engaged. The greater the ability to target material that will hold the user's attention, the greater the ability to sell more advertisements while also collecting more data.

Having siphoned away their users' private and personal information, the platforms first hoard the data and then harness it for their own purposes. In so doing, it becomes the entry drug to other online abuses.

REDEFINING PRIVACY FOR CORPORATE REWARD

"We have a responsibility to protect your information. If we can't, we don't deserve it," headlined a 2018 advertisement signed by Mark Zuckerberg.[26] The apology followed the revelation that Facebook allowed a third party—Cambridge Analytica—to gain access to the personal information of 50 million subscribers and use it for political manipulation.[27]

Protecting the data held about each individual is an important responsibility, but it is an after-the-fact result of the initial violation of an individual's privacy rights through data siphoning. In chapter 6 we saw how Facebook unilaterally redefined privacy expectations. As Mark Zuckerberg explained, "We decided these would be the social norms now, and we just went for it."[28]

The platforms' approach to privacy has taken many forms over the years. At first, they relied on the users' ignorance about the volume of private information that was being collected and how that collection was accomplished. As consumers became more aware, the companies switched their emphasis to their "privacy policies," misleading users that such policies were about protecting privacy when they were really about gaining permission to exploit user privacy. In today's redefinition, the message is "we care about your privacy," but "privacy" is defined as what happens to your personal information *after* it is taken. Such a defining of privacy as an after-the-fact "here's what you can know about what has already been taken" is a misrepresentation bordering on fraud.

The repositioning is exemplified by an advertising campaign Facebook ran to burnish its image by showing sensitivity to practices that concern consumers. In the campaign's privacy-focused ad, perky, soothing background music sets the scene for a Facebook employee named Rochelle to talk about her role on the "Facebook privacy team." After showing photos of her family, Rochelle explains, "I actually help people understand their privacy, because it means different things to different people." Then she describes what "privacy" means to Facebook: "You should be able to understand who has your data and how they use it."[29]

Rochelle's explanation is the equivalent of a bank robber assuring depositors he will take good care of and send reports about his ill-gotten gains. Focusing on how your personal information is used and what you can do to look at it obfuscates how the information was collected in the first place.

While Facebook has featured heavily in these pages, the practices being discussed—from the institutionalized invasion of privacy to the "we care" cover-up—are endemic across the major digital platform companies. When the *Washington Post* headlined "Amazon's Newest Products Expand Its Surveillance Inside the Home," the company dodged the issue in its response by rolling out the "we care" redefinition of privacy. "Privacy is foundational to everything we do," Amazon's spokesman reassuringly replied. "We continue to give customers greater transparency over control of their data."[30]

But allowing insight about what information has already been taken and how it is being used is not "privacy." *Webster's New World Dictionary* defines "privacy" as "The quality or condition of being private . . . one's private life or personal affairs."[31] The latest effort to redefine "privacy" as what happens after your private information has already been taken is mistaken at best and misleading on purpose. It has, however, become the go-to obfuscation.

Too often the major platforms define privacy as anything but the wholesale and ravenous taking of personal information and converting it into a corporate asset used to manipulate the very users from which it was originally purloined. Such an opportunity for manipulation will rise to new heights in the metaverse and the models driving AI. As discussed in chapters 4 and 5, the harvesting of personal information is expanding from clicks to biometric data to become the feedstock of "You ain't seen nothin' yet!" capabilities.

FIFTY IS TOO OLD

The new excitement about the metaverse and AI, however, should not obfuscate prioritizing the need to do something about the current privacy practices of the major platforms—practices that carry over to the onrushing new capabilities.

Once again, we as citizens of the early twenty-first century find our policies defined by decisions made in a different era to address different challenges. Fifty years ago, as relational databases developed, they introduced the worry that comparing different data files could reveal otherwise private information. As the then-largest collector of data, the federal government convened an advisory group to deal with "a growing concern that automated personal data systems present a serious potential for harmful consequences, including infringement of basic liberties."[32]

The result was a 1973 report to the Secretary of Health, Education, and Welfare. The report proposed "safeguards against its [data collection] potentially adverse effects." Those safeguards were a "Code of Fair Information Practice."[33] It is this half-century-old set of standards—commonly referred to as FIPs—that has informed the discussion of computer privacy ever since. The five FIPs are:

- There must be no personal data record-keeping systems whose very existence is secret.

- There must be a way for an individual to find out what information about him or her is in a record and how it is used.

- There must be a way for an individual to prevent information about him or her that was obtained for one purpose from being used or made available for other purposes without his consent.

- There must be a way for an individual to correct or amend a record of identifiable information about him or her.

- Any organization creating, maintaining, using, or disseminating records of identifiable personal data must assure the reliability of the data for their intended use and must take precautions to prevent misuse of the data.

As Professor Woodrow Hartzog observed in his wonderfully titled *The Inadequate, Invaluable Fair Information Practices*, "While the FIPs have been remarkably useful, they have painted us into a corner."[34] That corner is how the challenges of digital information have moved from the management of stored data to the practices surrounding how data are collected and capitalized. We are wrestling today with, in Shoshana Zuboff's powerful and accurate description, how digital platforms have become "an economic system built on the secret extraction and manipulation of human data."[35]

The arrival of computer chips in everything, coupled with ubiquitous wireless networks to connect those chips, has enabled business plans built on the aggressive acquisition, aggregation, and monetization of data far beyond anything imagined in the 1970s. The FIPs establish a baseline, but we cannot continue to remain in a fifty-year time warp. It is no longer adequate to rely on policies developed for another era.

THE INSUFFICIENCY OF CONSENT

The 1973 FIPs introduced the concept of "consent" into the digital information lexicon. "There must be a way for an individual to prevent information about him that was obtained for one purpose from being used or made available for other purposes without his consent," the principles recommend.[36] In the intervening decades, both platform companies and policymakers have embraced this concept of "notice and consent" to underpin their activities.

"Consent" may have been sufficient protection half a century ago when smaller, more focused amounts of data were collected. Today, however, "consent" is insufficient to address the needs created by the ubiquitous gathering and exploitation of personal information.

"Consent" is too often subverted by coercion—When the only way to receive a service is to click "I Agree" at the end of dense legalese, that is coercion, not consent. Such a demand approaches a "contract of

adhesion"—an agreement in which one party's outsized power allows it to dictate the terms to the other.[37] "I am holding your service hostage until you agree that I can invade your privacy" is not consent.

"Consent" puts the burden on the wrong party—Just because you didn't lock your car is not permission for it to be stolen. Just because you use an internet service should not be permission to pillage your personal information. The collection of personal information should impose special obligations on the collector, not the target. Requiring the object of a privacy intrusion to be responsible for the ongoing defense of their privacy is an upside-down expectation.

The con of "consent"—Alleging that "consent" is synonymous with "privacy protection" has been used to block meaningful privacy oversight. When platform companies argue the user is in control of their information, they are hiding behind the "user consented" fig leaf. Even if consumers can be considered to consent to collection, the use of that data to make decisions that end up affecting individual lives is far beyond the reach of "consent."

It's not about "consent," it's about power—"Privacy is about power, not consent (or harm)," Professor Lisa Austin has observed.[38] The obligation for the responsible exercise of power rests with the party that has the power. Basing privacy on consent of the individual insulates the powerful collector of personal information from that responsibility. Because the power of the dominant digital platforms comes from their use of private information, then the obligation for the responsible exercise of that power should shift to the companies.

"Consent" is not scalable for humans—As requests for the absolution of consent increased, consumers' ability for any meaningful review did not increase. There is an infinite corporate capability to expand the collection of private data, but a cognitive limit of humans' ability to deal with the constant deluge of pop-ups asking for consent. The result is "consent fatigue," which encourages users to simply agree in order to get on with what they want to do, thus further advantaging the companies.

"Consent" has discouraged the search for meaningful privacy protections—As if the abuses perpetuated in the name of consent are not bad enough, being able to hide behind "consent" has removed the incentive

to search for a twenty-first-century update to the FIPs. "Consent" has become the legal, political, and public relations blocking force to resist improved privacy expectations. How to reach the roots of the capitalization of personal information that is harnessed in harmful ways must come out from behind the "consent" smokescreen.

BUILDING PRIVACY BY DESIGN

Multiple times in this book I have cited Mark Zuckerberg's revealingly honest explanation of Facebook's 2009 decision to unilaterally create a new privacy policy: "We decided that these would be the social norms now and we just went for it."[39] It is a grade A example of innovators making the rules to benefit themselves.

Something as important as the privacy of personal information cannot be left to engineers and entrepreneurs to define on their own. "For centuries, we have accepted that there are some areas where private companies belong—and other where they don't," Azeem Azhar observed in *The Exponential Age*, "People's private selves . . . shouldn't be on the market."[40]

How then should we define and enforce privacy-protecting behaviors? As presented in the preceding chapter, implementing privacy by design can be accomplished without heavy-handed government micromanagement but through codes of conduct developed in a multistakeholder process and approved and enforced by government.

There must be an alternative to "We decided."

There must be privacy protections beyond the hollow concept of "consent."

There must be a reinstitution of the Duty of Care that has been steamrollered by the thrill of building something and then remaking the rules to fit the impact of its consequences.

There must be the ability to protect privacy early in the development process rather than having to fight to play catch-up later.

In 1995, Ann Cavoukian, the commissioner of Information and Privacy for the Canadian province of Ontario, proposed the concept of "privacy by design."[41] "[P]rivacy assurance must ideally become an organization's default mode of operation," she wrote.[42] Privacy by design is based on seven principles:[43]

- Proactive not reactive; preventative not remedial
- Privacy as the default setting
- Privacy embedded into design
- Full functionality—Positive-sum, not zero-sum
- End-to-end security—Full life cycle protection
- Visibility and transparency—Keep it open
- Respect for user privacy—Keep it user-centric

At the heart of privacy by design are the same concepts that underlie the common law Duty of Care: the anticipation and mitigation of privacy-invasive harms.

Privacy by design has been widely discussed in the intervening decades since its origination. The EU's General Data Protection Regulation (GDPR) references it in Article 25 of the rule.[44] The International Organization for Standardization (ISO) has had a seventeen-nation developmental process underway since 2018.[45] The U.S. Federal Trade Commission (FTC) included it as a part of the agency's 2012 recommendations on privacy.[46]

Moving beyond the 1995 principles of privacy by design to actual policy implementation has been a rocky road. Just as happens to every expansive status quo–challenging idea, a coterie of commentators has picked it apart. Among the criticisms are that software engineering is too complex,[47] or the principles are too vague.[48] While these may be relevant issues, eventually the question comes down to: compared to what? And the current meaningful abuse of proactive privacy protections.

The specifics of Cavoukian's principles are a roadmap for moving the rules for the present and future of online privacy out of the hands of those whose business benefits from the abuse of individual privacy. Privacy by design simply suggests that while determining how digital products function, it is of even greater importance to determine what the impact of those functions will be on personal privacy and how to mitigate any adverse effects.

How those who invade privacy continue to make the rules—and how standard codes can be a solution—is illustrated by the standard-setting activities of the companies that collect information from within our homes. In 2021 the Connectivity Standards Alliance (CSA) announced Matter, "the interoperable, secure connectivity standard for the future of the smart home."[49] Google, Amazon, Apple, Comcast, and almost 200 other companies are "participating in bringing the Matter specification, reference implementations, testing tools and certification programs to life."[50] The "key attributes" of the new standard are simplicity, interoperability, reliability, security, and flexibility.

Nowhere in the development of the standards for the products we invite into our homes is there discussion of controlling the amount of information the devices collect. Instead of meaningful privacy protections, the standard setters roll out the old "protect your information after we've taken it" ruse.[51]

The Matter standards-setting process demonstrates both the failure of the information-gathering companies to police themselves in regard to protecting users' privacy, as well as the structural opportunity such a process offers for the implementation of privacy by design concepts.

SHOOTING BEHIND THE DUCKS

If we are to get ahead of the ever-expanding collection and exploitation of personal information, it is going to require government intervention, as surely as government was involved to rebalance the industrial abuses of the Gilded Age.

Thus far policymakers have been "shooting behind the ducks" in their efforts to protect personal privacy. The need is for proactive guidance for the behavior of the platforms rather than sporadic, after-the-fact attempts to "fix" some of the effects of those behaviors. Getting in front of the challenge means establishing the expectation that privacy protection is designed into data collection technologies as a forethought rather than an afterthought and that such designs apply retroactively.

Chapter 9 discussed a new approach to oversight of the issue through the development of enforceable industry behavioral codes patterned after the multistakeholder technical standards–setting process. Engineers have demonstrated it is possible to successfully wade through the complexities of digital code to collaborate in the identification and mitigation of potential technical problems. The same can-do attitude and process should be applied to identify and mitigate privacy problems as well.

Industry-developed technical standards are implemented with the goal of increasing operational efficiencies and thus profit, whereas some fear the privacy protections could have the opposite effect. This is why it is essential that the government act. The continued failure of the companies to act in a meaningful manner to protect personal privacy has invited government, on behalf of the people, to expect more of them. Only government has the ability to identify the issues, assemble the affected public and private parties to develop a code of privacy expectations, and enforce those decisions.

The digital innovators were unafraid to step up and define behaviors for the new world they were creating. Neither should the representatives of We the People be inhibited from stepping up to protect individuals when those decisions tilt against the public interest.

CHAPTER 11

Competition by Design

COMPETITION HAS LONG BEEN THE AMERICAN IDEAL. YET THE INCEN-
tives of industrial capitalism, and now internet capitalism, frequently
push to constrain or even crush that ideal.

During the Gilded Age, companies such as Rockefeller's Standard
Oil delivered wonderful benefits to American consumers, just as the
digital platforms have done today. In the process, however, Standard Oil
and others like it engaged in predatory and anticompetitive practices.[1]

Today, governments at the state, federal, and international level are
investigating, litigating, and legislating to address the anticompetitive
activities of digital platform companies. The assertions made in these
actions echo the activities of the original Gilded Age: that the companies
have delivered many wondrous products and services for consumers, but
at the cost of snuffing out competition and causing consumer harm.

The fact that Standard Oil or the digital platform companies grew
large and profitable is not the problem. It is not against the law to be big.
The antitrust laws are not about size but rather about whether a company of any size behaves in a manner that makes it harder for others to
compete.

If a company can avoid competition, it can inoculate itself from having to focus on what is best for consumers as opposed to what is best for
the company. Eliminating competition also exterminates the engine that
drives innovation while permitting those energies to be directed toward
better ways to exploit consumers or expand market power into adjacent
areas.

A lack of meaningful digital platform competition is seminal in the abuses of the dominant digital platforms. In the preceding chapter, we saw how a lack of competitive alternatives forced consumers into a take-it-or-leave-it situation, resulting in their loss of privacy. In the next chapter we will see how the lack of competition denies consumers the ability to "change the channel" and thus permits platforms to control the information citizens receive.

In the original Gilded Age, the search for solutions to abusive corporate power organized into two camps: those pushing for a "break them up" kind of solution and those advocating a "regulate them" approach. The debate over platform power in the new Gilded Age often seems to be following the same lines. Reasserting a competitive marketplace today requires an "all of the above" solution. Antitrust enforcement is crucial, of that there can be no doubt. There are multiple competitive issues, however, that can best be addressed by agile regulatory oversight.

Antitrust enforcement is principally an after-the-fact action; even when used to review pending mergers, it is reactive to someone else's initiative. Regulation can be more proactive, including the opportunity to "design for"—this time, to design for marketplace behaviors that create rather than thwart competition.

TRUSTS AND ANTITRUST

In chapter 1 we saw how the idea of using a trust to control business activity was the brainchild of John D. Rockefeller's lawyer, Samuel Dodd, in 1882.[2] At the time, companies were incorporated by individual states and were not allowed to operate across state lines. Dodd's idea allowed the shareholders of the state-bounded companies to contribute their stock to a legal fiction, a corporate trust, that would manage the collective corporate affairs. In the case of the Standard Oil Trust, that management meant coordinated pricing, production, and other activities in order to dominate the market.

In 1889 the state of New Jersey, seeking revenue to pay off Civil War debts, changed its incorporation laws to allow activity across state borders, thus increasing the revenues of those companies.[3] Permitting the establishment of a company to hold the stocks of other companies eliminated the need for corporate trusts. The Standard Oil Trust became Standard Oil of New Jersey. The term "trust" stuck, however, as a synonym for a large, coordinated enterprise exercising undue market power.

Just as today, Gilded Age legislation to curtail the abusive practices of such enterprises was oft discussed but not acted upon. "Despite a century of congressional rumblings against monopolies, federal efforts to pass trust-busting legislation did not advance," modern competition policy leader Senator Amy Klobuchar recounts in *Antitrust: Taking on Monopoly Power from the Gilded Age to the Digital Age.*[4]

Finally, in 1890, the U.S. Congress passed the Sherman Antitrust Act.[5] The Sherman Act established a two-tier test to determine whether the activities of a dominant firm were anticompetitive.[6] First, it prohibited the use by multiple firms of their collective market power to control prices or exclude competition. Second, it prohibited single firms—monopolies—from engaging in exclusionary conduct such as predatory pricing, refusal to deal, or similar practices to achieve, maintain, or enhance that power.

Even with such concepts enshrined in statute, however, big business continued to behave pretty much as before passage of the Sherman Act. The years that immediately followed enactment saw mergers and acquisitions continue at a rapid pace. In 1899 over 1,200 companies were consolidated. In 1900 another 1,000 firms were rolled into one another.[7] The result was the proliferation of market-dominating entities, including the Sugar Trust, the Sewer Pipe Trust, even the Thread Trust.[8]

The prima facie evidence that the Sherman Act could be subverted came in 1901 when J. P. Morgan combined ten separate steel companies—including the largest, Carnegie Steel Company, and second largest, Federal Steel Company—into a new behemoth, United States Steel Company. The resulting giant controlled more than half the nation's production of steel and was the world's first company to be valued in excess of $1 billion.[9]

Laws may be on the books, but they are of no value absent their enforcement. President William McKinley refused to deploy the Sherman Act against U.S. Steel. Shortly after the consummation of the merger, McKinley feted its architect, J. P. Morgan, at the White House.[10]

William McKinley was assassinated in September 1901. Theodore Roosevelt became president of the United States, and the scales began to tip in the other direction.

NOT WHETHER BUT HOW

The late nineteenth century produced a "growing apprehension about the direction of capitalism," Bhu Srinivasan observed in *Americana: A 400-Year History of American Capitalism*.[11] In a description of the period that reminds us of the present, he explained, "Yes, industrial capitalism produced a consumer dividend that was plain for many to see in terms of choice, price, and quality. But the complexities of new products and new materials left the consumer, among others, increasingly vulnerable without some basic protections."[12]

In search of such protections, citizens began to organize. As a result, the last decade or so of the Gilded Age overlapped with what historians have named the Progressive Era. During this period the people's representatives in government began to assert oversight of the activities of big business. Theodore Roosevelt became their champion.

"At the start of Roosevelt's presidency in 1901, big business had been in the driver's seat," Doris Kearns Goodwin explained in *The Bully Pulpit*, "Under Roosevelt's Square Deal, the country had awakened to the need for government action to allay problems caused by industrialization."[13]

In 1906 Roosevelt's Department of Justice harnessed the Sherman Antitrust Act to file suit against Standard Oil of New Jersey. The government argued that Standard Oil was a conspiracy in restraint of trade and thus prohibited by the Sherman Act. Five years later, in May 1911, after working its way through the lower courts, the U.S. Supreme Court found for the government and assessed as a remedy that the giant be broken into thirty-four separate companies.[14]

While often labeled a "trustbuster," Roosevelt wrestled with the same kinds of issues that inform today's discussion of marketplace oversight: whether to rely on antitrust law or regulatory oversight. Goodwin succinctly describes the dilemma: "Roosevelt considered the Sherman Act a blunt instrument that might prove 'more dangerous for the patient than the disease.'"[15] Instead, he preferred "new legislation enabling the national government to examine corporate records and determine what remedies, if any, were needed."[16]

The politics of the time, however, were also a precursor of today as conservative Republicans opposed an expansion of federal power. "With the path to effective regulation blocked by a stubborn conservative Congress," historian George Mowry explained, "the only way for Roosevelt to bring the arrogant capitalists to heel was through the judicious use of the antitrust laws."[17] As a result, during the Roosevelt administration, the U.S. government brought forty-five antitrust suits against big businesses. Even more were pursued by his successor William Howard Taft, who brought an additional seventy-five actions.[18]

Roosevelt worried, however, that pursuing justice through the courts was insufficient because it was inefficient. Bringing legal action is a reliably lengthy process that promises an uncertain outcome. There was, he lamented, an "irksome and repeated delay before obtaining a final decision" of the courts such that "even a favorable decree may mean an empty victory."[19]

Roosevelt was also bothered by the burden that accompanied the simultaneous prosecution of multiple cases. In his 1907 Message to Congress, the president warned that "to attempt to control these corporations by lawsuits means to impose upon both the Department of Justice and the courts an impossible burden; it is not feasible to carry on more than a limited number of such suits."[20]

To big business and their allies in Congress, it was precisely these burdens that made the judicial system preferable to sector-specific oversight by a regulator with expertise in the field. That an alleged abuse had to exist before suit was brought was itself an invitation to roll the dice, allow the abusive behavior to proceed, and see what happened. Even if caught and an antitrust suit ensued, the offending activity could most likely continue through the multiyear term of the trial. In defense of its actions, the corporation had a rich supply of attorneys earning top dollar to compete with those representing the government. Even if the company lost, it faced the blunt remedies of antitrust law rather than the targeted focus of ongoing rules from an expert agency.

Roosevelt was a "trustbuster" because antitrust law was the tool that was available. But he had even greater faith in sector-specific regulation. In 1906, just before the filing of the antitrust suit against Standard Oil, Roosevelt offered the company a deal: in lieu of the lawsuit, Standard should accept detailed public oversight.[21] The company rejected the offer, but the philosophy behind it was clear: Roosevelt was not so much concerned about size but behavior, and he believed that behavior could be controlled by regulation. "The design [of government]," he explained in his 1907 Message to Congress, "should be to prevent the abuses incident to the creation of unhealthy and improper combinations, instead of waiting until they are in existence and then attempting to destroy them by civil or criminal proceedings."[22]

Theodore Roosevelt left office in 1909. When his hand-picked successor, William Howard Taft, ran for reelection in 1912, he was opposed not only by Democrat Woodrow Wilson but also by a disappointed Roosevelt, running as the candidate of the Progressive ("Bull Moose") Party. Roosevelt campaigned on what he called "New Nationalism," a vision first seen in his 1906 offer to Standard Oil: government would accept monopolies so long as it could control their activities through regulation.

Both Roosevelt and Taft lost to Woodrow Wilson. But the tide of public opinion and public policy had shifted. The question was no longer whether there should be oversight of big business but how that oversight was to be accomplished.

BEYOND THE SHERMAN ACT

On December 14, 1911, one day shy of seven months after the Supreme Court handed down its *Standard Oil* decision, the U.S. Senate Committee on Interstate Commerce convened to consider the future of antitrust policy. The first witness introduced himself simply: "My name is Louis D. Brandeis; I live in Boston and I am a lawyer by profession."[23] It was understatement in the first degree. Louis Brandeis was known as "The People's Lawyer" for leading the fight against what he labeled "The Curse of Bigness."[24]

Brandeis' testimony lasted three days. His message was as simple as his self-introduction: monopolies harmed democracy and individual opportunity, and existing antitrust laws and government structures were insufficient to deal with the threat.[25] As Brandeis scholar Jonathan Sallet observed, "The impact of his advocacy between 1911 and 1914 helped propel the enactment in 1914 of both the Federal Trade Commission Act ['FTC Act'] and the Clayton Act."[26] The FTC Act created a new federal antitrust enforcer to protect against not only "unfair methods of competition but also unfair acts or practices."[27] The Clayton Antitrust Act established in statute the illegality of activities that "may be to substantially lessen competition."[28]

For the next fifty years, Progressive Era concepts were at the heart of government oversight of the economy. Together, antitrust and regulatory enforcement combined to protect competition and marketplace dynamism.

THE COMPETITION LAW CHALLENGE TODAY

Competition law enforcement is the confluence of the decisions of presidents and judges. William McKinley's failure to act on the U.S. Steel behemoth demonstrated the role of politics and presidential whims. It falls to jurists, however, to interpret the broad instructions of both the Sherman Act[29] and the Clayton Antitrust Act.[30] Such decisions become problematic as judges wrestle, not only with the vagaries of the law, but also with rapidly changing technology. As antitrust scholar Herbert Hovenkamp has observed, judges have "the unusually difficult task of applying extremely open-ended statutory language to an exceptionally open-ended set of circumstances."[31]

The political and interpretative realities of antitrust law combined in the 1960s to lead the policy away from its roots. A new antitrust theory combined the laissez-faire politics of Reagan Republicanism with simplified economic analysis that helped judges narrow the open-endedness dilemma Hovenkamp described. Because the theory was incubated at the University of Chicago, it has become known as "Chicago School" jurisprudence. The theory holds that if left alone, the market will work things out, and thus the consequence of an antitrust decision that mistakenly identifies violations (a so-called "false positive") is greater than that of missing the infractions ("false negative").

The Chicago School theory was popularized in Robert Bork's 1978 book *The Antitrust Paradox*.[32] The "paradox" to which it referred asserted that antitrust enforcement, purportedly brought for the purpose of protecting consumers, actually raised prices to those consumers by protecting inefficient companies. The question in any merger, Bork argued, was whether it advanced consumer welfare, which, he asserted, was usually defined by lower prices. Rather than judging corporate behavior based on its impact on the marketplace, the so-called consumer welfare standard shifted the plaintiff's burden to establishing whether the behavior in question would raise prices for consumers.

The consumer welfare standard has become an ideological article of faith among conservative politicians and the jurists they appoint and confirm to the bench. A current majority of the U.S. Supreme Court appear to be advocates of the Chicago School.

Yet the practices of the dominant digital platform companies present new challenges for the consumer welfare standard. The Roosevelt administration, in making its decision to prosecute Standard Oil, had to choose between the lower prices the company offered and the deleterious effects of industry concentration Rockefeller had achieved. It is the same kind of consideration that must be weighed in the case of the digital platforms, where the profit-maximizing price is free. As former assistant attorney general Makan Delrahim has pointed out, "Price effects alone do not provide a complete picture of market dynamics, especially in digital markets in which the profit-maximizing price is zero."[33]

The evolution from an industrial economy to an information economy will of necessity force a reconsideration of the application of antitrust concepts in the new era. Such a relook has opened the door to a return to the earlier, more expansive approach some have labeled "neo-Brandeisian."[34] The antitrust theories of the Progressive Era were developed, after all, in response to the changes created by a new industrial economy. The changed circumstances created by the internet economy provide a similar impetus today.

Our challenge today is to discover a new way of understanding market power that reflects the new nature of the marketplace created by digital assets. As Jonathan Sallet observed in *Louis Brandeis: A Man for This Season*, "We are living in times that Louis Brandeis would have understood."[35]

ANTITRUST: AN INCOMPLETE SOLUTION

Antitrust laws are sweepingly generic while also being assiduously specific. They focus on the generally applicable issue of corporate concentration on a case-by-case, company-by-company basis. Without a doubt, antitrust laws—and their enforcement—are critically important to protecting a competitive marketplace, yet they are but one part of the solution.

The remedy in big antitrust cases such as the 1911 Supreme Court decision in *Standard Oil Co. of New Jersey v. United States*[36] or the 1982 consent decree in *United States v. American Telephone & Telegraph Co.*[37] was to break up the offending company into smaller parts. The response to the dominant digital platforms, many believe, should be a similar "break 'em up" solution.

The deployment of antitrust enforcement in the twenty-first century is challenged by the difference between industrial assets and digital assets.

Breaking up Standard Oil into thirty-four different companies or AT&T into seven "Baby Bells" was possible because of the physical facilities involved. Standard Oil of Ohio and Southwestern Bell were capable of being cleaved off to survive on a freestanding basis because they were facilities-based enterprises. Breaking up one of the dominant digital platforms would be more challenging since the most important assets are virtual instead of physical.

The source of the platforms' dominance is data collected from consumers and squirreled away behind a protective moat in order to control competition. If, for instance, Facebook (Meta) was to be disassembled, the surviving mini-Metas would need to be self-sufficient in a world where self-sufficiency is defined by access to data. Giving multiple mini-Metas such a data trove may create potential competitors, but also runs the risk of recreating in the hands of a few companies the kind of data dominance that caused the anticompetitive behavior in the first place.

Brute "break 'em up" antitrust enforcement also creates what public interest advocate Harold Feld calls the "starfish problem." Analogizing to the world of marine invertebrates, Feld warns, "Certain species of starfish have tremendous powers of regeneration. If you tear one piece, the individual pieces grow into a new starfish." Thus, "when separating the component pieces of dominant digital platforms, it is entirely possible that each segment will become dominant in its own line of business."[38]

The other challenge created by reliance on the antitrust laws is the simple matter of how long an antitrust case takes and how fast technology advances. The last two major antitrust cases, *United States v. American Telephone & Telegraph Co.* and *United States v. Microsoft Corp.*, each took more than ten years from opening investigation to conclusion.[39] During this lengthy period, the companies being prosecuted were allowed to continue the activities for which they were being sued, as well as use the profits from those activities "to entrench themselves and protect their position."[40]

Of even greater importance when it comes to antitrust remedies is how in today's world of fast-paced technological developments the world changes during the pendency of the antitrust action and how those changes have no bearing on the decision. The decade required for the AT&T and Microsoft cases to be resolved is a lifetime in tech development. Consider, for instance, a few of the technologies that became mainstream in the decade leading up to the publication of this book: augmented reality, artificial intelligence, the internet of things, touchscreens, cloud computing, and robots, to name just a few.[41] The reliably lengthy period required for an antitrust suit means that by the time the litigation works its way through the many levels of the judicial process, the technology and marketplace that existed at the time the suit was filed can be nothing more than a memory.

These forces create "what may be the most significant issue surrounding antitrust enforcement against major digital platforms—identifying proposed remedies that will reliably be more beneficial than harmful," observed Phil Verveer, the lead attorney at the start of the AT&T case. He calls this "an essential consideration in determining whether to bring suit, and one likely to require as much analysis as the facts and circumstances that would support a finding of liability."[42]

Reliance on antitrust alone cannot be the only tool in the pro-competition toolbox. There is also a need for competition promoting rules broadly applicable to an industry, not just a company, as well as the ability to move ex ante with regulation, not just ex post through litigation. The speed of technological change also establishes a need to move with relative dispatch as opposed to the years-long process of antitrust litigation.

Beyond the functional realities of antitrust enforcement there is a limit on the kinds of abuses it can reach. It is outside of plausible antitrust action to address how the platforms purloin consumer information almost at will. It is similarly beyond the scope of antitrust laws to address the quality of information being distributed by the platforms. It is hard to shoehorn these issues into either the Sherman Act's "monopolize, or attempt to monopolize" or the Clayton Antitrust Act's "substantially harm or lessen competition" tests.

Antitrust economist Carl Shapiro explained the importance of regulation as well as litigation this way: "Antitrust is not designed or equipped to deal with many of the major social and political problems associated with the tech titans, including threats to consumer privacy and data security, or with the spread of hateful speech and fake news. . . . Addressing these major problems requires sector-specific regulation."[43]

Once again, we return to Theodore Roosevelt. In his 1907 Message to Congress, he noted: "That centralization in business has already come and can not be avoided or undone."[44] As a result, "The public at large can only protect itself from certain evil effects . . . by providing better methods for the exercise of control through the authority already centralized in the National Government by the Constitution itself."[45]

To Roosevelt, it was a question of sovereignty. Just who was sovereign, the companies or the people acting through their government? "What is needed," he told Congress, "is not sweeping prohibition of every arrangement, good or bad, which may tend to restrict competition, but such adequate supervision and regulation as will prevent any restriction of competition from being to the detriment of the public."[46]

OPENING WHAT IS CLOSED

The concept of "competition" is more than "anti-monopoly." A goal of public policy should be the promotion of competitive behavior, not just redress in the case of its absence. To accomplish this necessitates regulatory policies oriented toward identifying the problem and getting in front of it. Chapter 9 discussed a process for accomplishing such oversight through the development of enforceable codes of conduct for digital platforms.

The development of a competition-promoting code does not mean the imposition of micromanaged "utility regulation" that dictates how the platforms will operate. Rather, it means the identification of risks and the development of a code of practices to deal with those risks.

One of the seminal risks to competition in today's online platform world is the closed structure that the companies have designed to deny others the ability to compete. When you sign up for Facebook, YouTube, Twitter, or other platforms, you are locked into that platform, unable to seamlessly relate its activities with other platforms and able to communicate only with other users on that platform. Opening up that construction by insisting platforms are interoperable—like the internet itself—would free consumers to be able to make competitive decisions and eliminate monopoly control of the information taken from those consumers.

Some of the smaller online platforms have embraced such interoperability through a so-called fediverse, a federated network of servers. Their activities have demonstrated that interoperability is not constrained by technical issues. Instead of being isolated behind a corporate wall, these platforms interconnect so that their users can communicate regardless of which platform they choose. Doing this means, for example, that users are no longer trapped by the policies of the platform their friends are on but can select among the privacy or content policies of platforms with which they are the most comfortable while still being able to share with their friends.

Interoperability has been dubbed the competition "super tool" by Yale economist Fiona Scott Morton.[47] Mandating that the platform's operational protocols, as well as its data hoard, interconnect with other services lowers the barriers to competition without imposing the problems that come from the detailed intervention of industrial era regulation.

The key concept advanced by interoperability is that of freeing the consumer so that there can be competition *in* the market. Competition *for* the market has already been defined by network effects and the barrier to entry they create. Interoperability is a simple policy that frees users to exercise competitive choice while leaving scale and network effects in place.

Industrial era rules required regulators to "make choices that come with the risk of creating inefficiencies" that in the digital era are "heightened because of the rapid change of products and prices over relatively short periods of time," Scott Morton explained.[48] Interoperability, in contrast, is light-touch oversight that only requires the establishment of an interface, not the dictation of products or management of the company. In particular, as proposed in chapter 9, when the interface is designed through a multistakeholder process, the regulator can focus on oversight of the process itself to ensure it meets the identified goals, followed by the acceptance and enforcement of the outcome to ensure it promotes competition.

Freeing Consumers through Interoperability

Contrary to the capabilities of the internet, the platforms have locked up their users. The internet was originally called "internetworking" because its standards made possible the interoperability of diverse networks. Popular services such as email take advantage of such interoperability to build platforms that work with each other. A Gmail message can reach a customer with an Outlook email address thanks to their designed-in interoperability. The noninteroperable design of the dominant digital platforms, however, are the equivalent of a Gmail account being unable to reach an email account on a different service.

The platforms have exploited the internet's interoperability when it suits their purposes, such as the ability to reach across services to collect information about users. But the platforms have designed their key consumer-facing functions to be noninteroperable in order to lock up those users. Such practices are not only an affront to the public interest, they are also a perversion of the openness of the internet that allowed these companies to be created in the first place.

Denial of interoperability as a path to market dominance is not new, nor is its remediation. The telephone monopoly that was broken up in *United States v. American Telephone & Telegraph Co.* was assembled in part by denying small local phone companies the ability to interconnect to AT&T's long-distance network unless they sold out to AT&T or operated on terms the bigger company dictated. To oversee the fair and efficient functioning of interoperable telephone networks, the FCC developed detailed policies to mandate and manage that interconnection. Such a heavy-handed approach should not be necessary in a digital platform market where an interoperability code of conduct has been adopted.

By shutting down interoperability, the platforms shut down consumer choice. This means, for instance, there is little choice for consumers when it comes to protecting their personal privacy. As network effects drive the consumer to a specific platform, that platform can force the user into a privacy take-it-or-leave-it simply because there is no alternative platform with equivalent connections to friends and business associates. Interoperability would mean there could be a social media platform for privacy-conscious users that would still be able to communicate with their friends on a different platform.

Interoperability is about freeing consumers to make choices. It is not about imposing identical platform functions. Platform security and data protection, for instance, would continue to be determined by the companies themselves (and could become competitive positioning). Similarly, the interconnecting platforms would be free to develop their own systems and services. Interoperability simply tears down the walls that today prohibit consumers from having a meaningful choice.

Interoperability is common in the software world. Unlike how platforms keep their users in a walled garden, software relies on the ability of different programs to interoperate. This is accomplished through what is called an application programming interface (API). When, for instance, your food delivery app displays a map tracking the order's progress, that map most likely does not belong to the delivery app but is licensed from a third party and interfaces with the delivery app through an API.

Platforms use APIs as well, for instance to collect information about you from other sites. But the platforms specifically eschew APIs that might allow users to escape to choose among alternatives.

In the early days of mobile text messaging, I experienced the default instinct against interconnection. As originally implemented in the United States, the exchange of short messages was—just as on today's digital platforms—possible only within the same cellular network. If you were on AT&T, for instance, and wanted to send a text to a Verizon subscriber, you were out of luck. As the head of the wireless industry association, I proposed the establishment of a clearinghouse to allow all messages to flow across all carriers. The initiative was shot down because one company believed that trapping their customers this way would force the friends of those customers to sign up with the company as well. Fortunately, interconnection ultimately prevailed. As a result, text messaging exploded—and so did internetwork competition.

One must ask, why does interoperability work for text messaging but not for platforms? Why does interoperability work for email but not for dominant platforms? Why is it impossible for a video posted on YouTube not to be posted on a Facebook timeline?[49] Why cannot a consumer shop for acceptable privacy protections from among interoperating platforms?

The lack of interoperability by the dominant digital platforms is simply the latest iteration of the anticompetitive incentive of companies to control their users. It is a problem that has been confronted and remediated before—and can be again.

Opening Monopoly Control of Data through Interoperability

It is not only their users that the platforms hold hostage but also the personal information collected from those users. In this regard, the companies create a new monopoly over the essential asset of the twenty-first century.

As previously discussed, the capital asset of the twenty-first century is data. Because the data collected about individuals are the currency of the platform economy, whoever has the most data can drive market concentration by denying the data to others. When the platforms collect the personal information of users and hoard it, they are creating a monopoly as surely as when John D. Rockefeller collected and monopolized the output of oil wells. Because of the soft nature of digital assets, however, remediation of such data domination practices can be addressed by interoperability and interconnection.

As we have seen, a barrel of oil is "rivalrous" in that using it denies its use to anyone else, while software is "nonrivalrous." A data file can be used repeatedly by multiple parties without diminishing its functionality. Interconnecting nonrivalrous data assets would allow competition to flourish. The platforms, however, have chosen to behave in a manner that monopolizes the data assets so as to dominate the market.

The success of a government-mandated and government-supervised, industry-developed data interconnection program to open competition has been demonstrated in the United Kingdom on financial platforms. The UK's Open Banking Initiative,[50] ordered by the Competition and Markets Authority (CMA) regulator, required the nation's nine largest financial institutions to, upon a consumer's request, make available to third parties the data the banks possess about that consumer.[51] Financial institutions, like the digital platforms, had traditionally used their control of such information to lock consumers into their relationship with the bank and cripple competitors.

The purpose of the Open Banking order was to increase competition in financial services by allowing consumers to request the data the banks held about them be shared with competitive services, including smaller banks and innovative online services. The big banks still held the consumer's data, but at the consumer's request had to make it available to others.

The program is overseen by the Open Banking Implementation Entity (OBIE), created and funded by the banks and overseen by a CMA-appointed trustee. The OBIE established standards for mandatory APIs to allow different entities to securely access and interface with the banks' databases. The APIs became widely usable in the summer of 2019. As of early 2022 there were approximately 245 authorized third-party providers using the APIs to offer competitive services.[52] Participating banks had grown from the original mandatory nine to over ninety, as even smaller banks that are not required to participate do so for their own competitive reasons.

Making interoperable the data financial institutions hold on their users has proved so successful that fifty countries around the world are now exploring similar concepts.[53]

Some platforms have attempted to define the issue of data interoperability in terms of "data portability." Mark Zuckerberg, for instance, wrote in the *Washington Post*, "Regulation should guarantee the principle of data portability. . . . If you share data with one service, you should be able to move it to another."[54] It is a replay of the linguistic jujitsu that redefined privacy.

Data portability may sound like data interoperability, but it is not. The key word in the Zuckerberg endorsement is "share." "If you share data . . . you should be able to move it" is not the same as "the data held about you should be interoperable with other services." Sharing describes the ability to move photographs, music, or other files stored on one service (i.e., "shared") to another.[55] It is helpful, but it is not the open interoperability of data assets necessary for algorithms to be competitive.

As we have seen, the fediverse companies have demonstrated the viability of interoperability. So have the dominant digital platforms in other contexts. Google, for instance, offers a service for Google Cloud users that allows them to share data.[56] Amazon calls its similar service "Amazon Clean Rooms."[57] As Google explained, "Rising to meet today's business challenges can require companies to collaborate . . . with outside organizations, and across geographies, while pooling and enriching joint data sets."[58] Similarly, Amazon explains how the availability of Clean Rooms "helps companies across industries easily and securely analyze and collaborate on their combined data sets."[59]

Because data are fungible and nonrivalrous, it can be exchanged using secure methods. The kind of supervised code process discussed in chapter 9 could certainly be put to work to bring competition to online platforms through interoperability processes that, as the Google Cloud announcement proclaims, "help unlock the value of secure data collaboration."

TEAR DOWN THIS WALL!

The major platform companies have achieved their dominance by building a wall to shut out the inherent interconnectedness of the internet. It is a reality that is broadly beyond the scope of antitrust enforcement. Regulatory oversight can thus help competition in a manner that is additive to antitrust enforcement. Successfully applied, its ex ante actions can concurrently even mitigate the need for ex post antitrust enforcement.

The nature of such oversight should be to encourage competition by design. The multistakeholder process described in chapter 9 can be the vehicle for establishing such light-touch expectations of responsible behavior. The continued failure of the dominant digital platforms to emerge from their walled gardens and embrace competition has invited government to stand for competition and choice.

CHAPTER 12

Truth and Trust by Design

PROFITING FROM UNTRUTH IS AN AMERICAN TRADITION. AT THE heart of the delivery of news has always been a commercial incentive; and nothing sells like sensationalism and outrage—regardless of its veracity.

Such a commercial incentive remains unchanged in the twenty-first century. What has changed is the unprecedented power over the flow of news and information possessed by a handful of dominant social media platforms and the failure of those companies to embrace meaningful journalistic standards. The ability of AI to create false images and audio only adds to the problem.

The publication of outrage—especially political outrage—is a well-worn business strategy dating back to the time of the Founding Fathers. "The golden age of America's founding was also the gutter age of American reporting,"[1] Eric Burns observed in *Infamous Scribblers: The Founding Fathers and the Rowdy Beginnings of American Journalism.* Early American news was a business "to be molded into whatever shape he [the publisher] thought would be most pleasing to his customers and thus most profitable to him."[2]

In the Gilded Age news continued to be a business shaped by publishers. Like today, that business was transformed by technology. The steam-powered rotary printing press churned out newspapers at an unprecedented rate, printing on both sides, folding, and cutting with high-speed efficiency that reduced the cost to deliver information and extended the newspapers' reach.[3] One effect of the new technology, Paul

Starr noted in *The Creation of the Media*, was that as publishers "expanded the scope of the news, they also expanded journalism beyond news, publishing outright hoaxes" in an effort to gain circulation and revenue.[4]

It would seem the more things change, the more they stay the same. The internet has created the greatest imaginable opportunity for the dissemination of information and ideas. On top of the internet, however, the platform companies have perfected profiting from the dissemination of misinformation, disinformation, and malinformation. In the process, the old adage that the remedy for bad information is more information has been destroyed. When profits are in the targeted delivery of what people want to see, "more information" does not mean diversity of information, but simply more of the same.

TECHNOLOGICAL TRIBALISM

As we have seen, the dominant digital platforms operate on a common business model: the collection of data about personal preferences and the use of that data to both dominate the market and to pair the user with specific information. The previous chapters looked at designs to protect privacy and promote competition in such an environment. This chapter looks at how the business model reinforces tribal instincts, including how it can affect the democratic process.

In the early days of the internet, it was anticipated the network would create a new digital public commons for the sharing of ideas and opinion. It was hoped that when anyone with an internet connection had the potential reach of the *New York Times* or NBC, the diversity of voices would be a boon to democratic ideals.

But then the social media algorithms stepped in.

These algorithms maximize revenue for the platforms by denying citizens two essential components of the free flow of information. The first to suffer is the quality of the information as social media platforms curate for cash. Whether the information selected for distribution is spreading lies, hate, or the propaganda of adversarial nations is secondary to whether it maximizes the number of advertisements the user sees. The second negative effect to the free flow of information is how the platforms lock in the user, force feeding them the selected information without the ability to "change the channel" and discover countervailing information.

The algorithm-driven consequences of social media platforms are far different from the effects of an algorithm that recommends a book or answers a search query. By selecting specific types of information, often of questionable origin and veracity, to micro-target individuals, platform algorithms corrupt the ability to overcome tribal instincts through an understanding of the broad public good.

The founding fathers chose as the national motto *E pluribus unum*—"Out of many, one." To create such a *unum* requires an electorate informed by commonly shared information—including information the individual citizen may not seek out on their own. The business plan of social media platforms works to contradict this democratic imperative.

Instead of providing a common set of facts, social media algorithms are programmed to deliver content designed to agitate and stimulate tribal prejudices. It is a highly profitable business strategy created to hold the users' attention to generate revenue by displaying as many advertisements as possible. By segregating information in this manner, the social media platforms quarantine audiences into divergent realities.

The platforms profit from turning the founders' concept upside down to emphasize *pluribus* over *unum*.

To achieve these results, the social media companies harness four interrelated techniques: audience targeting, curation for amplification and engagement, treating news as if it were an advertisement, and secrecy.

THE TARGET ON YOUR BACK

For as long as there has been the distribution of advertising messages the goal has been to target the information to where it could achieve the greatest efficiency.

It doesn't make much sense to advertise beef to vegetarians or gutter cleaning to those who don't own homes. Advertisers have long tried to focus their expenditures on a delivery vehicle likely to appeal to the designated audience. Manipulators of public opinion, for whom the first rule is "go to your base," have sought similar precision. Social media have made such targeting a precise process.

Gilded Age retail pioneer John Wannamaker is reputed to have lamented, "Half the money I spend on advertising is wasted; the trouble is, I don't know which half."[5] Wannamaker's concerns were a driving force in American information dissemination for most of the twentieth century. While targeted publications such as *Sports Illustrated*, *Jet*, or *Better Homes and Gardens*, along with broadcast sports and soap operas, helped broadly identify certain types of audiences, mass communications typically meant a common message delivered to masses of people. Vegetarians saw the ad for the steak special at the grocery and apartment dwellers saw ads for gutter cleaning.

When, on June 1, 1980, in Atlanta, Georgia, Ted Turner inaugurated Cable News Network (CNN), few appreciated how it was the beginning of a process that would culminate in the targeted news distribution of digital platforms. I was invited to speak at the CNN launch ceremonies. Struggling to articulate the groundbreaking nature of how new television networks would be segmented much like specialty magazines, I described CNN as a "telepublishing event marking a watershed in information provision."[6] It was a bit over the top rhetorically, but the metaphor stands: the targeted diversity of the newsstand had come to dedicated video channels. Cable television overcame how the limited number of airwaves constrained broadcast channels to open a plethora of new channels that parsed the mass audience into self-identifying targets. MTV was for teens, BET for African Americans, Nickelodeon for kids, and the Golf Channel for . . . well . . . golfers.

It was in the news business that the impact of cable's audience targeting began to affect democratic political discourse. For much of the twentieth century, journalistic commentary was supposed to be separate from reporting of the news and reporters were expected to turn out opinion-free products. This held to the old business assumption that balanced, down-the-middle reporting maximized the number of users and thus maximized advertising revenues. The introduction of cable television news channels changed that tradition as the competitive need to distinguish among outlets made opinion and ideology a product differentiator.

The difference between the resulting cable news silos and the targeting of social media platforms is choice. The cable viewer can choose among multiple viewpoints at will—even flipping back and forth for the broadest collection of information. Social media platforms offer no such choice. When you sign on to a social media service you see what the algorithm selects for you to see and the ability to get a second opinion is limited.

Ted Turner ushered in a revolution in information delivery that enriched democracy by expanding citizen choice in access to news. The social media platforms have combined precise user information, limitless delivery capacity, and powerful targeting algorithms to enrich themselves by constricting citizen access to democracy's essential ingredient.

Playing for Amplification

In the Gilded Age the ability to amplify news for profit had a deleterious effect on the truth. "Faking was a rampant journalistic practice during the final quarter of the nineteenth century," one study of journalistic practices concluded.[7]

Gilded Age newspaper publishers "would do anything to attract audiences, up to and including printing rumors, fake news, and wildly sensationalized articles which helped incite a war with Spain," Jonathan Rauch observed in his excellent *The Constitution of Knowledge: A Defense of Truth*.[8] Press magnates such as William Randolph Hearst[9] and Joseph Pulitzer[10] relished the power to select and then amplify the news and voraciously exploited those abilities for both profit and political influence.

So widespread was the practice of amplifying questionable news for commercial gain that it was given a name: "yellow journalism."[11] Among the characteristics of yellow journalism, one authoritative analysis concluded, was "faked interviews and stories, misleading heads [headlines], and a parade of pseudo-science."[12]

As the Gilded Age wound down in the early twentieth century the journalists themselves began to push back against the commercial and/or political excesses of the press barons.[13] The 1922 founding of the American Society of Newspaper Editors brought forth the following year a code for fact-based editorial behavior.

The first item of the editors' code, in capital letters, was "RESPON-SIBILITY" and the obligation of "considerations of the public welfare."

A subsequent canon, "SINCERITY, TRUTHFULNESS, ACCU-RACY," provided: "By every consideration of good faith a newspaper is constrained to be truthful. It is not to be excused for lack of thoroughness or accuracy within its control." The final canon, "DECENCY," held: "A newspaper cannot escape conviction of insincerity if while professing high moral purpose it supplies incentives to base conduct."[14]

For the remainder of the twentieth century, at least for credible news outlets, amplification power was offset by editorial curation responsibilities. Editorial opinions existed, but making the judgement about the source, truthfulness, and context of news information prior to its media magnification was the first job of editors.

For social media platforms, however, curation for quality has too often been a quaint historical concept. Editorial code-based judgment by professionals has been replaced by computing machines programmed to amplify rather than assess a piece of information. As we saw in chapter 3, corporate chieftains can decide to program the algorithms to favor credible news—as Facebook did briefly leading up to the 2020 election—or to prioritize for the increased revenue-producing engagement that comes from agitation-producing content—as Facebook returned to after the election.

Today's social media platforms have traded curation for accuracy for curation for amplification, engagement—and cash.

Treating News as Advertising

Amplifying for engagement has resulted in news and information being treated as a commodity similar to advertising. The result, Jonathan Rauch points out, "pretty much guaranteed an attention-seeking race to the bottom."[15]

News has always been the bait to attract readers/viewers so they could be exposed to advertisements. Social media companies have used their ability to deliver targeted news to create a hospitable environment for specific targeted advertisements.

The force behind the what-news-to-send decision is not the credibility of the piece of information, but rather whether it fits the profile of what will hold the recipient's interest long enough to deliver more advertisements. Hearst and Pulitzer manipulated the news to drive broad-based subscriber growth to stimulate advertising revenue; social media algorithms manipulate the news to keep users engaged with their platforms for the same purpose.

The psychological techniques used by slot machines to keep users feeding in money were studied and applied by social media companies.[16] Slot machines hold the user's interest by teasing them with small payouts; social media companies hold attention by teasing their emotions. Content that arouses the user is most likely to engage that person's attention and keep them connected to the platform.

The detailed knowledge of the prejudices and preferences of each user allows the platforms to manipulate citizens through the news they distribute. Facebook investor Roger McNamee explained the phenomenon this way: "When we check a news feed, we are playing multidimensional chess against massive artificial intelligences that have nearly perfect information about us. The goal of the AI is to figure out which content will keep each of us highly engaged and monetizable."[17]

One is far more likely to stay online when the content delivered is that with which they are comfortable—and the platforms have the personal information to know those predilections. The result is a "filter bubble" where the news the algorithms select is chosen to reinforce or push emotional buttons more than to inform. The same techniques that improve advertising efficiency end up producing a democratic inefficiency that creates and then reinforces different perceptions of reality.

"We [social media] have definitely helped to create these isolated chambers of thought and it is because of the mechanics of how the system works," Twitter founder Jack Dorsey told an interviewer.[18] Those mechanics, because they are designed to appeal to tribalism—and the ability to associate online with people in the tribe—are the antithesis of the search for common understanding that is necessary to achieve common good.

DECISIONS IN DARKNESS
Working together, targeting, amplification, and news prioritization have a force multiplier effect on the outcomes they produce. An equally destructive practice is how the force multiplier operates in secret.

One form of this secrecy is to hide what is being delivered. When the *New York Times* or Fox News distributes their content, everyone knows that it is happening and what that news is. When a social media platform such as Facebook or YouTube distributes a piece of news, it happens in the darkness of an algorithm and is known only to those targeted by the algorithm.

The inability to know something is being distributed, and to whom, cripples the ability to catch and repair lies.

Platform secrecy also hides editorial content decisions. When the *Wall Street Journal* or MSNBC makes an editorial decision about how to cover a piece of news the decision is obvious. When a social media algorithm makes an editorial decision, it is hidden. The inability to know the editorial prejudice behind a piece of information denies the consumer of the information a key basis for making a judgment.

Keeping up with such secret decisions, let alone keeping track of them, is beyond the linear capabilities of humans. Even the algorithm programmers cannot keep track of what is happening in their own software as it makes decisions in milliseconds. That the programmers who built the algorithms cannot keep up with the decisions made by their creations seems like dystopian science fiction. The resulting impact, however, is very real world as secrecy preempts accountability. When the absolute decisions—and absolute secrecy—of computers replace the nuanced analytical decision-making of humans, technology has taken us into the unknown.

"Informed public opinion is the most potent of all restraints upon misgovernment," the U.S. Supreme Court held in 1936.[19] The soft underbelly of "informed public opinion" is not only targeting tribalism, but also doing it in the dark.

DESIGNING FOR TRUTH

The excesses of information manipulation are neither new nor unresolvable. Such resolution, however, is more difficult when dealing with speech-related issues than for other platform created challenges such as privacy or competition. The protections of the First Amendment to the U.S. Constitution constrains government action that could limit expression. Yet the story of yellow journalism—both its technological stimulation and the solutions that developed—can be informative to our experience today.

Chapter 9 laid out a process for the development of codes of conduct for digital platforms. It is a model that could be applied to address some of the harmful information-based practices of the dominant digital platforms. There are two specific actions that lend themselves to such a process: a reprise of the 1923 newspaper editors' code, and the ending of the secrecy that shrouds the flow of information.

Reprising Newspaper Editors' Code of Conduct

Print moguls like Hearst and Pulitzer may have been powerful media potentates, but that did not stop others—even those in their employ—from collectively developing a common code of editorial conduct. The same should be true in the age of powerful platform moguls.

Thus far, the social media platforms have failed to come up with a meaningful equivalent to the 1923 American Society of Newspaper Editors code. Perhaps this is because while the editors may have worked for the moguls, they were individuals with a conscience that enabled them to challenge the bosses. The algorithms that make digital editorial decisions have no conscience and thus pose no challenge to the absolute rule of their corporate masters.

It must be possible in the digital era to establish standards of professional reference like those that helped harness the excesses of yellow journalism in the analog era. Absent, however, the conscience of human newspaper editors there is need for another enabling mechanism.

Since the social media platforms appear unable to develop meaningful editorial standards by themselves, government encouragement is required. Yet the line between such encouragement and a First Amendment violation is exceedingly fine. That such a fine line has been drawn previously and upheld by the U.S. Supreme Court suggests that an effort to find a First Amendment–respecting solution could be possible.

In 1949 the FCC implemented the fairness doctrine,[20] a rule that required broadcasters to air contrasting points of view on topics of public importance. The Supreme Court decided[21] the First Amendment did not prohibit the fairness doctrine.[22] The court's decision is inapplicable to online platforms, however, because it was based on the "unique physical limitations of the broadcast medium" (airwave scarcity) that established broadcasters as gatekeepers to publicly owned airwaves. Since the public, through the FCC, grants broadcasters the right to the limited asset of the airwaves, the decision held, the public has a right to assure their asset was used fairly.[23]

The establishment of behavioral norms like the code adopted by newspaper editors is one way of confronting current platform practices while respecting the First Amendment. Determining the exact role of the federal government in such a process is the key to addressing potential First Amendment concerns.

The seriousness of the misinformation issue should stimulate the search for a way to make an enforceable behavioral code work. The government may not (and should not) be able to actively influence decisions about lawful speech, but—without engaging in content decisions—there must be a way for We the People, acting through our government, to encourage the platforms' better angels to develop oversight within the protections of the First Amendment.

One example of such an indirect role for government was the Code of Practices for Television Broadcasters developed by the National Association of Broadcasters (NAB) in the 1950s.[24] The Television Code went far beyond the fairness doctrine to explicitly describe the kind of content that could be broadcast, including prohibiting profanity, disrespect of religion, and banning the positive depiction of illicit sex, drunkenness, and addiction. The code was voluntary, yet there was a quiet awareness that ultimately, when the FCC decided whether to renew the station's license, it would be important to demonstrate good character worthy of such a public trust. Failure to follow the code's definition of responsible behavior could create an opportunity to challenge the character of the licensee and thus threaten renewal of the valuable license. After operating for almost thirty years (1952–1983), the NAB code was dissolved, not because of First Amendment issues, but because of antitrust concerns.[25]

Absent some authoritative incentive, however, it appears that the dominant platform companies will not develop a code that could define responsible behavior.

One such nudge—which would require an act of Congress—could be to condition the liability protections the platforms enjoy under Section 230 of the Communications Act on a company's adherence to an industry code. Since Section 230 protects platforms from being held liable for content they disribute, adherence to a code of conduct could be the safe harbor.

In *Regulating Digital Industries*, Mark MacCarthy developed a thoughtful and intricate First Amendment–protecting proposal.[26] His idea would require the social media companies to abide by rules of transparency and accountability that would be developed in cooperation with industry and enforced by a federal regulatory commission. To deal with the First Amendment issues, MacCarthy would delegate the resolution of disputes arising from the code to an industry self-regulatory organization.

The United Kingdom's oversight policy might suggest another First Amendment-respecting approach. In that country, the Office of Communications (OFCOM) is responsible for the oversight of the networks, much as the FCC does in the United States. Unlike the FCC, however, OFCOM is not a part of the parlimentary majority's government, but an independent board that selects the agency's leadership and oversees its activities. Whether such a body with delegated governmental authority, but not directly a part of government, would withstand First Amendment scrutiny is problematic, but at least worthy of scrutiny.

Alternatively, the Federal Trade Commission (FTC) could possibly rule that a company's failure to implement a behavioral code is a violation of the prohibitions against unfair acts or practices under the FTC Act. Once such a code is included in the company's terms and conditions it constitutes an agreement with the customer that can be enforced by the FTC as unfair or deceptive if not followed (much as the FTC did with its action against Facebook for the latter not following its own privacy code). This would substitute consideration of whether the companies were adhering to the agreement they made with customers for a government ruling on editorial content.

The conscience of journalists helped to overcome the excesses of yellow journalism. The substitution of soulless algorithms for the conscience of human editors makes such independent action more difficult today. The effect of social media's anti-*unum* and often fact-free activities is of such magnitude, however, to challenge us all to think creatively about a First Amendment–respecting analog to the editorial standards of the mainstream media.

Ending Secrecy

One First Amendment–safe place for such a code to start is to shine light on the secrecy that shrouds social media's selection and dissemination of news and information. Transparency has always been the sentinel that guards the truth. Yet the business plan of social media platforms denies users the transparent information necessary to make judgments about what they are receiving.

If the platforms are going to outsource content curation to their users, then the new curators need to know about the veracity and context of the content they receive—facts the platforms heretofore have refused to release. Transparency also requires knowledge that there is information being distributed.

The failure of social media platforms to be transparent about what they distribute and even whether they are distributing it is at the heart of their information abuses. A code of platform accountability could simply provide that the companies allow light into the process and give consumers access to the information necessary to make an informed decision.

Federal policy should encourage the companies to develop such a code. Whether the code is overseen by a new digital agency, or comes into being as a condition precedent for the liability protections of Section 230, or is an effort by the FTC to protect against unfair or deceptive acts, or exists as a voluntary and enforceable decision of the companies themselves, consumers should be given the information with which to make informed judgments about what they are being fed by social media platforms.

Fortunately, the digital technology that created the platforms also provides the tools for transparency and truth. The same technology that has been programmed to exploit can be programmed to protect—easily, at minimal cost, and without affecting the platforms' free speech rights.

Such a solution can be accomplished through an application programming interface (API), the capability that allows different software programs to work together.[27] In this instance, the platforms would provide a "public interest API" to the input and output of the algorithm. The functioning of the algorithm—a protected editorial determination—would not be touched. Using such a public interest API, however, third parties could build their own algorithms to understand what today is secret: the source and delivery of the information. Such information is openly available today in traditional media but kept secret by social media.

Such a public interest API could be designed to not threaten the intellectual property in the algorithm, the editorial rights of the platform, or the privacy of personal information. Public interest APIs would simply create the opportunity to make the same kind of assessment that is made every day about traditional media but denied to social media users: what information is distributed and what are the facts involved.

Media watchdog groups on both the left and right of the political spectrum have long been able to assess and report on the editorial decisions of newspapers and television. Public interest APIs would extend that capability to social media by giving the watchdogs the tools to understand what goes into and comes out of the algorithmic black boxes.

Because such open access is not government interfering with editorial decisions, it creates less of a First Amendment challenge. Yet at the same time it provides the tools to understand what is going on under the First Amendment protections. Most of all, it provides to consumers—whom the platforms have forced to be the new information curators—the tools necessary to make the decisions that have been imposed on them.

WHAT ONLINE PLATFORMS OWE DEMOCRACY

"All the armies of Europe, Asia, and Africa combined . . . could not take a drink from the Ohio or make a track on the Blue Ridge in a trial of a thousand years," Abraham Lincoln counseled. "If destruction be our lot we must ourselves be its author and finisher."[28]

Social media platforms have been the beneficiaries of American democratic capitalism. They owe it to that democracy to be its protector, not its finisher.

As Americans, we should fear the potential abuse of government power over the flow of ideas. We should also fear the failure to deal with the actual ongoing abuse of social media platforms that is gnawing away at American democracy.

The power of the digital platforms to regulate speech far exceeds that of the government. The failure to responsibly exercise that power has created significant problems for the successful operation of a democracy. There is more at stake than the quarterly financial results of the social media companies.

The First Amendment is sacrosanct, yet in the past we have been able to respect free speech principles while also expecting the dominant provider to behave responsibly.

We would be wise to listen to another of Abraham Lincoln's admonitions: "As our case is new, so must we think anew, and act anew."[29]

Part V

Consequences We Control

"*The real problem of humanity is the following: we have Paleolithic emotions, medieval institutions, and god-like technology.*"

Edward O. Wilson[1]

Chapter 13

Time to Make History Again

PULLING IT ALL TOGETHER

The stories of the original Gilded Age recount the unprecedented solutions developed to protect the public interest in the midst of the unparalleled changes of industrialization. The scope and scale of the industrial economy made resolving those challenges a step into the unknown. Embracing the need to act, the citizens of the era secured their place in history by establishing structures that stood the test of time for the next century.

We find ourselves in a similar position today. The unprecedented capabilities of digital technology have created the need for us to once again act in a new manner to assert societal standards.

Thus far, the history of twenty-first century digital platforms has been composed by those who write software code and their investors. The actions or non-actions we take to establish rules to restore the primacy of the public interest as opposed to corporate interests will be the legacy of our times.

Today's experience is the result of digital shapeshifting. The world of physical communications networks has been transformed by a new superstructure that replaces the physical with the virtual. New digital upheavals ranging from the metaverse to generative AI will advance that virtualization to become immersively all-encompassing.

The digital world will only become more complex. Absent intervention, the dominant digital platforms will only become more controlling.

That the metaverse and generative AI are already upon us before we have successfully dealt with the challenges created by the early iterations of the platform economy should be a loud alarm bell summoning us to action.

The new digital reality calls for more than tinkering around the edges. Our actions must be as bold and innovative as the ideas that produced the platforms themselves. Pruning the digital platforms' activities here and there may feel like doing something, but an occasional trim does not go to the root of the matter.

To get to that root requires new tools. The oversight model built to protect the public interest in the industrial era is inadequate to meet the pervasive power of the dominant digital platforms.

In the original Gilded Age, when confronted by never-before-seen industrial challenges, the people, acting through their government, developed never-before-seen solutions. They invented an antitrust statute in 1890. They invented the first independent regulatory agency in 1887. The people's representatives protected the supply of food and drugs, the safety of workers, and so much more that was in the public interest. Confronted by our own never-before-seen digital challenges, We the People and our representatives must be equally creative and bold.

It is our turn to step up and make history.

RULES AND REFEREES

Because digital innovation, with all its imperfections, is the key to economic growth, it is essential we find a new oversight model that encourages innovation while protecting consumers and competition. In the original Gilded Age, regulation mirrored the operating practices of the companies it oversaw. It is a formula worth replicating today. Digital public policy should mimic the forces it oversees by itself being agile and responsive.

The fact the old regulatory model is struggling in these new times is not an argument against regulation, but rather a demonstration of the need for a new regulatory paradigm to deal with the new digital realities. We must bring to the development of a new oversight model the same vigor and vision as the platforms themselves brought to create the new challenges in the first place.

It is simply illogical that the companies whose activities drive national economic growth while at the same time shaping individual lives would exist without societal expectations for the betterment of the public good.

The plan for governing the ungoverned begins with addressing how the digital companies manage the technology they develop. While the companies may lobby against rules to protect average consumers, they embrace rules when they themselves are the consumer. The technology the digital platforms employ is rigidly rules-based, only the rules are called "standards." Such standards—collectively developed by multiple stakeholders ranging from software coders to hardware manufacturers and the platforms themselves—assure the quality of functional results and allow the technology to interoperate.

Standards allow the technology marketplace to function. Standards can similarly allow the consumer marketplace to function.

Replicating the process that created technology standards to develop consumer-facing behavioral standards is the twenty-first-century equivalent of regulation that "looks like your pet." As previously discussed, industrial oversight looked like the management model of the industrial corporations it regulated because legislators and regulators simply imported to their activities the "scientific management" that Frederick W. Taylor was preaching to the companies. If industrial Taylorism held that production efficiency resulted from quashing individual initiative in favor of enforced rules, the thinking went, the same must hold true for the oversight of those industrial activities.

The rigidity of such regulatory micromanagement is not applicable in today's innovative and fast-changing digital economy. Digital companies have already replaced rigid industrial management with agile management and soon AI-aided management. Just as Taylorism migrated from corporations to the government, the anti-Taylorism of agile digital management, and ultimately AI-aided oversight, must as well. The role of government in the digital era should not be to micromanage, but to identify risk and establish behavioral expectations to mitigate those risks—and then to enforce whether those policies are implemented in the ever-changing technology and marketplace environment. Industrial regulation by dictate fits neither with the agile management and the incentive to innovate that defines the digital age, or with the machine cognition of AI.

The establishment of behavioral standards utilizing on a supervisory basis the kind of multistakeholder structure that has worked so successfully for technical standards can combine protection of the public interest with the agility necessary to keep up with innovative technological advances.

But there needs to be a referee to make certain the new rules are followed. In the example of technical standard setting, the enforcer is the companies. When it comes to behavioral standards regarding the use of that technology, the people's representatives in government must become the referee.

"If the internet is the most powerful and pervasive platform in the history of the planet . . . there needs to be a referee," I told the 2015 Mobile World Congress shortly after enactment of the FCC's net neutrality rules.[1] The job of government oversight in the digital era, I argued, was to act less as a regulator that micromanages and more like a referee with the ability to throw the flag to enforce behavioral expectations.

In the 2015 net neutrality rules, for instance, the FCC established that networks should be open and nondiscriminatory but did not dictate the operational decisions of the companies. Instead, the FCC established as a behavioral standard that the actions must be "just and reasonable." It was a policy reflecting the evolution from a regulatory dictator to a referee making calls based on well-established expectations.

The evolution from "the world's largest ungoverned space" to responsive and responsible regulatory oversight begins with a common set of rules. Because the United States has not pursued the development of such policies, however, those rules are being written elsewhere.

As we saw in chapter 8, the digital rules developed by the European Union address the same kinds of issues that have been repeatedly raised in hearings before the Congress of the United States. Protecting user privacy, preventing platforms from self-preferencing their own services, opening the control of app stores over what they make available, giving consumers control over their apps, and other market adjustments have all been well discussed in Congress. While Congress talked, the EU acted.

The interconnection of the internet means these rules—developed by other nations in good faith, but some argue, with a protectionist bent—will apply de facto to American companies and American consumers. It is important for both the consumers and creators of digital services that Congress establish American policy in regard to the issues of the digital era.

What It Takes to Make American Rules

America's policy mechanism is faltering when it comes to dealing with the challenges of digital technology.

Congress is in gridlock with performance politics, political vendetta, and prodigious lobbying too often trumping meaningful policy development.

The ability of regulators to establish new behavioral rules is constrained by statutes written in a different era to deal with different problems created by different assets and different technologies.

The judicial system's antitrust decision-making, while important, is reliably lengthy, ultimately uncertain, and not comprehensive.

The Washington truism that it is easier to kill something than pass something is alive and well today. Yet so was it alive when the Sherman Antitrust Act was passed, the Interstate Commerce Commission (ICC) was created, and so many other landmark industrial era statutes went on the books.

If our review of the original Gilded Age teaches anything, it is that persistence pays; the importance of advocating for new policies sets a policy agenda for years to come. It was twenty years from the founding of the Grange in 1867, to advocate for farmers against railroads, to the creation of the ICC. It was almost another twenty years until in 1906 the agency was given the necessary regulatory teeth.

Today, however, the long arc of history is being condensed by the speed of digital change. The almost forty years between organizing to oversee the activities of the dominant network of the industrial era and its meaningful regulation was literally a human lifetime in the nineteenth century.[2] Today, the lifetime for new technology is measured in a handful of years. The previous pace of technological development and adoption created a time buffer that allowed a more leisurely approach to policy development. Such a cushion exists no more. The technology of the new Gilded Age demands a more expeditious and agile timetable for the implementation of an overall regulatory plan.

Three times I have been involved in congressional development and enactment of legislation to establish rules to govern the activities of corporations that were taking advantage of new technologies. The Cable Act of 1984 brought cable television under federal regulation. In 1993 the same was done for the new cellular telephone business. The Telecommunications Act of 1996 then updated the Communications Act of 1934 to reflect the many other changes that had occurred in technology and the marketplace. In each instance there were three key factors necessary for the legislation to pass: the industry had to feel the pressure of external forces; there had to be a "get" for both the consumer and industry sides of the issue; and the Congress needed to delegate ongoing oversight to an expert agency. The same legislative forces are at play today in determining whether there will be rules for the dominant digital platforms.

External Forces

The business plans of the platform companies are built to profitably exploit the ubiquitous nature of the internet. As we have seen, the linear economics of the industrial era have been replaced by exponential internet economics. At almost zero marginal cost, the whole world has simultaneously become both a factory producing content and a marketplace for that content.

American companies' ability to continue to uniformly "build-once-sell-everywhere" is threatened if the internet is broken into regulatory fiefdoms. The absence of American policy leadership has exacerbated this potential. Politico's reporting is worth repeating: "A common refrain among European officials is that they're being forced to take action because the U.S. hasn't."[3]

We have seen how Meta Platform's 2021 filing with the Securities and Exchange Commission (SEC) starkly identified how external policy forces from abroad could threaten the platform business model. "If policy agreements cannot be reached," the 10-K report said, "we will likely be unable to offer a number of our most significant products and services, including Facebook and Instagram, in Europe."[4] It is hard to have such "policy agreements," however, when there is no American policy established by the United States Congress.

It is not just the actions of foreign nations that are creating external threats to the platforms' business plans. American federalism reserves to the states the right to operate in areas unregulated by the national government. The failure of the U.S. government to act has created a vacuum that individual states are filling with their own laws. Because the internet is oblivious to state boundaries, action by an individual state can create de facto national policy. Even more destructive for the companies, however, is how the lack of uniformity among state laws imposes significant burdens and uncertainties.

When California adopted the California Consumer Privacy Act (CCPA) in 2018, the decision affected platform behavior in the other forty-nine states. At the same time, however, the politics of some of those other states encouraged their legislators to act as well. Multiple states acting in their own unique manner creates a situation where even a slight difference in the statutes could result in purposeless diversity that harms the ability of the platforms to enjoy a contiguous, uniform domestic market.

The diminution of the internet's exponential economics as a result of divergent regulatory schemes at the state and nation-state levels is an increasing threat to the businesses the platform companies have built. The solution to the diversity among states is a preemptive federal law. The solution to multinational policy conflicts is more complex, but it begins with having American policies to put on the diplomatic table.

The external forces of conflicting policies should create the economic incentive for the digital platform companies to come out from behind their opposition to meaningful government oversight of digital markets.

Something for Everyone

The existential threat to the economics of the platform companies should open the door to the kind of trade-offs necessary to pass legislation. The companies now have a new economic calculus when it comes to determining their position on legislative action.

It was formerly in the companies' interest to oppose any legislation. When the calculus was the cost of regulation versus the cost of no regulation, nothing always won. This formula, however, no longer holds. The calculus now begins, not with nothing, but with the threat of an increasingly fractured market. The advantage of getting to the economic benefits of a common set of rules can now outweigh what once were the benefits of no rules at all.

This new calculus creates an opening for negotiations between the advocates for regulatory oversight and the platform companies. The position of the advocates of public interest protections with regard to privacy, competition, and misinformation has remained basically unchanged and is well known. The companies now have the incentive to engage in negotiations that would trade modification of some of their practices for regulatory uniformity.

The door to a "get" by both sides has swung open. The process of reaching an accord both sides can live with will be arduous. There is, however, now a reason for all sides to move from their policy mountaintops to the pragmatic development of rules for the internet era.

The thing that corporate chieftains hate the most is uncertainty. The certainty that they were the ones setting the behavioral rules for the marketplace was previously a great advantage to the companies. Now that is being replaced by the threat of uncertainty and confusion resulting from incompatible domestic and international regulatory policies. By their willingness to engage in the development of new policies, the platforms can create their own new certainty.

None of the parties to the policy negotiations will come away with everything they want. But all can get something they need: protections against harms for consumers, competition—and regulatory certainty.

Delegating to an Expert Agency
The legislative process is not optimized for detailed decision-making. Congress paints in broad brushstrokes to establish principles and then relies on others to fill in the detail. Such a process is both good politics and a wise and beneficial strategy.

It is good politics since the ability to pass legislation depends on 60 percent of the Senate and a majority of the House of Representatives supporting the action. It is simple human dynamics that agreement on broad principles is more attainable than addressing each and every issue in specific. Of necessity, Congress must deal in the broad scope of policy decisions rather than attempt to encode every little detail in legislative language.

It is also good policy to establish a set of underlying principles rather than cement specific details into the black letter law. Especially today, the pace of change demands an agile regulatory body acting under congressional principles to make judgments to align ever-evolving circumstances with the statutory guidance.

This is not a revolutionary concept. Governance of the financial markets, for instance, is based on broad principles implemented by the Securities and Exchange Commission (SEC). The same holds for the oversight of our food and pharmaceuticals, communications networks, automobile safety, and many other marketplace segments. Congress declares what it expects and then delegates to an expert agency the detailed day-to-day implementation of those policies.

Finally, it is an advantage to have an expert agency assigned to deal with the detailed implementation of the congressional instructions. While digital policy is an important issue, it is but one of a myriad of diverse issues with which Congress must deal. The job of a federal legislator is almost impossible in the scope of its expectations. On any one day it can mean dealing with issues of national security, tax policy, health care, infrastructure, the courts, and a seemingly limitless list of other issues. Members of Congress and the Senate, of necessity, must be generalists determining the four corners of a policy and delegating to an expert agency the ongoing oversight and implementation of that policy.

RULES FOR A VIRUS

In 2013, Google's executive chairman wrote, "Modern technology platforms, such as Google, Facebook, Amazon and Apple are even more powerful than people realize. . . . Almost nothing short of a biological virus can spread as quickly, efficiently or aggressively as these new technology platforms."[5]

The virus metaphor, no doubt benign when it was written in 2013, has a different image today. The pandemic occurred because the Coronavirus was uncontrolled. One might even say that the virus was making its own rules until reined in by individual behavior modification, national policies, and new vaccines. If the modern technology platforms mimic such a virus, there must also be similar antidotes.

Like a mutating virus, today's digital experiences are not an end state; they are but the current point on an evolutionary timeline. The virus of ever-evolving technologies will require ever-evolving vigilance. Dealing with the digital virus can follow a similar intervention plan as that utilized for the Coronavirus:

Testing—Any intervention begins with understanding the problem. We know the virus's behavioral infections: invasion of personal privacy, exploitation of the resulting data to concentrate markets, and control the flow of information. What we also need to understand are the techniques by which the platforms produce those effects. Such an understanding begins with transparency. Today the virus operates inside a software black box. Insight into the curation algorithms and content moderation decisions would allow an understanding of how the virus works.

Inoculation—This is a new virus; the inoculations that worked on previous infections are no longer adequate. As described in chapters 10 through 12, inoculation requires a strategy of ex ante behavioral designs to protect consumers and competition: privacy by design, competition by design, and truth by design.

Prevention—The virus will spawn new variants. There must be a permanent structure and ongoing oversight to deal with the new challenges. Chapter 9 described a new regulatory agency and a new regulatory paradigm to replace the rigid micromanagement of the industrial era. Ongoing digital oversight begins with agile regulation, based on a common set of principles, that is ever evolving to keep pace with the variants.

The beauty of capitalism is that companies evolve to meet changes in technology, markets, and regulation. When tech investor Peter Thiel warns, "We are in a deadly race between politics and technology," he is leading the tech companies down a false path.[6] First of all, what is dismissed as "politics" is not a handful of people in Washington, but the polity—the people—who are upset and looking for answers. Embracing policy solutions to practical realities is an opportunity for the platform companies. It is in the companies' interest not to slam the door in the face of regulation, but to be at the table as those policies are created. There must be—and will be—rules for the dominant digital platform companies. The only question is where those rules will be made.

History determines the greatness of a nation by how it deals with the challenges it confronted. In the original Gilded Age, the struggle was to reassert the public interest over corporate interest. The resulting framework drove the fruits of industrial capitalism to new heights to overcome depression, win wars, and produce an expansive economy that embraced competition and protected consumers and workers. A similar challenge with similar solutions is ours today in the new Gilded Age.

It is time to make history again.

Acknowledgments

Much of the original effort in this book was done in conjunction with my work at the Harvard Kennedy School and Brookings Institution. I am grateful to Nicco Mele for inviting me to become a fellow at the Kennedy School's Shorenstein Center, and for his initial challenge to use that time to address platform issues. Nancy Gibbs, who took over as director of the Shorenstein Center, expanded that opportunity. John Haigh, co-director of the Mossavar-Rhamani Center at the Kennedy School, then developed a creative program to take these ideas forward.

At the Brookings Institution Darrell West, vice president and director of Governance Studies, helped develop both programs and publications to further expand the effort. Darrell's support and suggestions to this work have been substantial and invaluable. Bill Finan and Yelba Quinn of Brookings Institution Press gave the book a home. Yelba, in particular, was a counselor, cheerleader, and shrink without whose continued support this book would not exist.

The idea for Part III, "Who Makes the Rules (in the Digital Gilded Age?)," germinated as a result of Kathleen Hall Jamison's invitation to deliver the 2017 Annenberg Lecture at the University of Pennsylvania.

Much of what is contained herein I owe to men with whom it has been my privilege to associate for years: Phil Verveer, Gene Kimmelman, and Jon Sallet. For over four decades I have been blessed to rely on Phil Verveer for guidance on public policy issues and an education in the law. A great public servant in multiple roles, Phil has always imparted a quiet wisdom and unerring guidance. Gene Kimmelman is one of the nation's foremost public interest advocates. Over the years, Gene and I have been both opponents and allies—it is much more fun to be on Gene's side. Jon

Sallet I first met when he was part of the Clinton administration, and I feared having to engage with his intellect. Fortunately, that relationship expanded and grew as Jon became both a colleague and friend during the Obama administration.

The amazing Elizabeth "Lizzy" Bateman was my research assistant. A student at The Ohio State University John Glenn College, Lizzy became my go-to for tracking down the untrackable. Watching Lizzy convinced me that the next generation's future is in good hands.

At the eleventh hour, Marjorie Pannell stepped in, reached above and beyond, and edited the final text. This is my second book working with Marjorie, and both the books and I have benefited greatly from her skill.

Multiple others contributed in many diverse and helpful ways. Imran Gulamhuseinwala helped me understand the UK's Open Banking Initiative. Mark MacCarthy—whose own book, *Regulating Digital Industries*, takes a deeper look at many similar issues—helped me think through some of the issues.

First and foremost, however, is Carol Wheeler. Somehow, Carol has found the patience not only to put up with me for almost forty years but to do so during the writing of multiple books. It is worth reiterating here what I have written previously: without Carol's love and support my life would be resoundingly less rich and I most certainly would be a different person. She has blessed me, our kids—and now our grandkids—with her love and wisdom. She is the shining light in my life and the ultimate supporter when her somewhat crazy husband announces he is again going to write a book. Carol, I love you.

Finally, Carol and I were blessed to call Mark Shields a friend. Mark's admonition, "Every one of us has been warmed by the fires we did not build, and every one of us has drunk from wells we did not dig," is a reminder of my gratitude to the many who welcomed me to their fires and wells to mentor and shaped my life, including Chuck Wheeler, R. M. Edgar, Wendell Ellenwood, Jacob Davis, George Koch, Julian Scheer, Michael DiSalle, and Bob Schmidt.

NOTES

PREFACE

1. John Steele Gordon, "Regulators Take On Silicon Valley, as They Did Earlier Innovators," *Wall Street Journal*, April 17, 2018, https://www.wsj.com/articles/regulators-take-on-silicon-valley-as-they-did-earlier-innovators-1524005477.

2. Mark Twain and Charles Dudley Warner, *The Gilded Age: A Tale of Today*, 1873, https://www.gutenberg.org/files/3178/3178-h/3178-h.htm.

3. Cat Zakrzewski, "The Technology 202: The Tech Industry Faced a Reckoning. Now Comes the Hard Part," *Washington Post*, July 22, 2021, https://www.washingtonpost.com/politics/2021/07/22/technology-202-tech-industry-faced-reckoning-now-comes-hard-part/.

4. Richard Edelman, "2021 Edelman Trust Barometer: Trust in Technology," March 30, 2021, https://www.edelman.com/trust/2021-trust-barometer/trust-technology.

5. European Commissioner for Internal Market Thierry Breton has warned that digital platform companies have become "too big to care" and "tend to neglect the consequences of their actions in our daily environments, whether it be in work [or] in our social lives, but also when it comes to our democracy." J. D. Kim, "Too Big to Care," *Technology Law Watch*, October 16, 2020, https://technologylawwatch.com/2020/10/16/too-big-to-care/.

6. Nicholas Davis, "What Is the Fourth Industrial Revolution?," *World Economic Forum*, January 19, 2016, https://www.weforum.org/agenda/2016/01/what-is-the-fourth-industrial-revolution/.

7. Anchal Gupta, "How Do Platform Businesses Win?," *Medium*, February 10, 2019, https://anchalgupt25.medium.com/why-do-platform-businesses-win-46485be8863b.

8. Zuckerberg quoted in Gupta.

9. "Putin says the nation that leads in AI 'will be the ruler of the world,'" *The Verge*, September 4, 2017, https://www.theverge.com/2017/9/4/16251226/russia-ai-putin-rule-the-world.

10. Susan Berfield, *The Hour of Fate: Theodore Roosevelt, J. P. Morgan and the Battle to Transform American Capitalism* (New York: Bloomsbury, 2020), p. 126.

11. Shoshana Zuboff, *Surveillance Capitalism: The Fight for a Human Future at the New Frontier of Power* (New York: Public Affairs, 2019).

12. Marko Saric, "How to Limit Your Exposure to the Surveillance Capitalism," *Medium*, May 30, 2019, https://medium.com/swlh/surveillance-capitalism-9294fc3a7709.

13. Nick Saint, "Eric Schmidt: Google's Policy Is to 'Get Right Up to the Creepy Line and Not Cross It,'" *Business Insider*, October 1, 2010, https://www.businessinsider.com/eric-schmidt-googles-policy-is-to-get-right-up-to-the-creepy-line-and-not-cross-it-2010-10.

14. Jonathan Rauch, *The Constitution of Knowledge: A Defense of Truth* (Washington, DC: Brookings Institution Press, 2021), p. 122.

CHAPTER 1

1. Naomi R. Lamoreaux, "Entrepreneurship in the United States, 1865–1920," in *The Invention of Enterprise: Entrepreneurship from Ancient Mesopotamia to Modern Times*, edited by David S. Landes, Joel Mokyr, and William J. Baumol (Princeton, NJ: Princeton University Press, 2012), p. 367.

2. Bernard Bailyn et al., *The Great Republic: A History of the American People* (Lexington, MA: D. C. Heath, 1985), p. 569.

3. Robert G. Barrows, "Urbanizing America," in *The Gilded Age: Perspectives on the Origins of Modern America*, edited by Charles Calhoun (Lanham, MD: Rowman & Littlefield, 2007), p. 102.

4. Andrew Noymer, "U.S. Life Expectancy," figure, faculty paper, Department of Demography, University of California, Berkeley, https://u.demog.berkeley.edu/~andrew/1918/figure2.html.

5. Sharon Basaraba, "A Guide to Longevity throughout History," Verywell Health, February 12, 2018, www.verywell.com/longevity-throughout-history-2224054.

6. Inaugural Address of Theodore Roosevelt, March 4, 1905, http://avalon.law.yale.edu/20th_century/troos.asp.

7. Ibid.

8. Ibid.

9. Ibid.

10. Jacques Barzun, *From Dawn to Decadence: 500 Years of Western Cultural Life, 1500 to the Present* (New York: HarperCollins, 2000), p. 539.

11. Marc Andreessen, "Why Software Is Eating the World," *Wall Street Journal*, August 20, 2011, https://www.wsj.com/articles/SB10001424053111903480904576512250915629460.

12. "Meritocracy in America: Ever Higher Society, Ever Harder to Ascend," Special Report, *Economist*, January 1, 2005.

13. Emmanuel Saez, "Striking It Richer: The Evolution of Top Incomes in the United States," faculty paper, Department of Economics, University of California, Berkeley, March 2, 2019.

14. Joseph Stiglitz, "The American Economy Is Rigged," *Scientific American*, November 1, 2018, pp. 57–61, http://www.scientificamerican.com/article/the-american-economy-is-rigged/.

15. Tim Wu, *The Curse of Bigness: Antitrust in the New Gilded Age* (New York: Columbia Global Reports, 2018), p. 59.

16. Scott Carpenter, "Steve Ballmer Becomes Ninth Member of the $100 Billion Club," *Bloomberg Wealth*, July 8, 2021, https://www.bloomberg.com/news/articles/2021

-07-07/steve-ballmer-becomes-ninth-member-of-the-100-billion-club#:~:text=One
%20of%20the%20world's%20most,to%20reach%20that%20lofty%20plateau.

17. "President Biden Address to Congress," C-SPAN, YouTube (video), April 28, 2021, www.youtube.com/watch?v=G9HLaxzdlwE.

18. Amy Klobuchar, *Antitrust: Taking on Monopoly Power from the Gilded Age to the Digital Age* (New York: Alfred A. Knopf, 2021), p. 73.

19. Ibid., p. 100.

20. StatCounter Global Stats, *Search Engine Marketshare Worldwide* (database), November 2022, https://gs.statcounter.com/search-engine-market-share.

21. Nicole Farley, "Search Advertising Is Thriving amid Economic Uncertainty, New Report Confirms," Search Engine Land, September 16, 2022, https://searchengineland .com/new-report-confirms-search-advertising-is-thriving-amid-economic-uncertainty -388004.

22. Diana Moss, "The Record of Weak U.S. Merger Enforcement in Big Tech," American Antitrust Institute, July 8, 2019, p. 5, www.antitrustinstitute.org/wp-content/uploads /2019/07/Merger-Enforcement_Big-Tech_7.8.19.pdf.

23. *Crunchbase* (database), https://www.crunchbase.com/search/acquisitions/field/ organizations/num_acquisitions/facebook.

24. Microacquire, https://acquiredby.co/amazon-acquisitions/.

25. Cited in *Federal Trade Commission v. Facebook, Inc.*, Case 1:20-cv-03590-JEB, August 9, 2021, p. 21.

26. George Miller Beard, *American Nervousness: Its Causes and Consequences* (New York: G. P. Putnam's Sons, 1881), p. vi.

27. Frederic Lardinois, "Google Now Has 1B Active Monthly Android Users," TechCrunch, June 25, 2014, https://techcrunch.com/2014/06/25/google-now-has-1b-active -android-users/?_ga=2.37656227.269269079.1632284460-964715258.1632185581.

28. White House, "Inaugural Address by President Joseph R. Biden, Jr.," WH.gov. www.whitehouse.gov/briefing-room/speeches-remarks/2021/01/20/inaugural-address -by-president-joseph-r-biden-jr/.

29. Ryan Barwick, "Nearly Half of All Ads on Fake News Sites Come from Google, Study Finds," Marketing Brew, June 16, 2021, https://www.marketingbrew.com/ stories/2021/06/16/nearly-half-ads-fake-news-sites-come-google-study-finds.

30. Jack Beatty, *Age of Betrayal: The Triumph of Money in America, 1865–1900* (New York: Vintage Books, 2007), p. 221.

31. Thomas Mann and Norman Ornstein, *The Broken Branch: How Congress Is Failing America and How to Get It Back on Track* (Oxford, UK: Oxford University Press, 2006), p. 7.

32. Richard White, *Railroaded: The Transcontinentals and the Making of Modern America* (W. W. Norton & Company, 2011), p. 22.

33. Tom Wheeler, *Mr. Lincoln's T-Mails: The Untold Story of How Abraham Lincoln Used the Telegraph to Win the Civil War* (New York: HarperCollins, 2006), pp. 19–21.

34. Central Pacific Railroad Photographic History Museum, "Pacific Telegraph Act of 1860," http://cprr.org/Museum/Pacific_Telegraph_Act_1860.html.

35. See RAND Corporation, "Paul Baran and the Origins of the Internet" (www.rand .org/about/history/baran.html); also see Tom Wheeler, *From Gutenberg to Google: The History of Our Future* (Washington, DC: Brookings Institution Press, 2019), p. 12.

36. Barry M. Leiner et al., "A Brief History of the Internet," The Internet Society, 1997 https://www.internetsociety.org/internet/history-internet/brief-history-internet/.

37. National Science Foundation, "On the Origins of Google," August 17, 2004, https: //www.nsf.gov/discoveries/disc_summ.jsp?cntn_id=100660.

38. AT&T 1956 Consent Decree, United States Congress Senate Committee on the Judiciary Subcommittee on Antitrust and Monopoly Antitrust Subcommittee, 1958.

39. Phil Goldstein, "How the Government Helped Spur the Microchip Industry," *FedTech*, September 11, 2018, https://fedtechmagazine.com/article/2018/09/how -government-helped-spur-microchip-industry.

40. National Science Foundation, "NSF and the Birth of the Internet," https://www.nsf .gov/news/special_reports/nsf-net/textonly/60s.jsp).

41. Wheeler, *From Gutenberg to Google*, pp. 162–163.

42. Remarks of Tom Wheeler, *Democratic Platform Drafting Committee, Day 3*, C-SPAN (video), July 29, 2012, at 27:43, www.c-span.org/video/?307328-1/democratic -platform-drafting-committee-day-3.

43. Joseph Keppler, *The Bosses of the Senate* (cartoon), *Puck*, 1889, https://brewminate .com/politics-and-corruption-in-the-gilded-age-1865-1900/.

44. E. R. Baker, *The Money Monopoly* (Des Moines IA: G. A. Miller, 1892), p. 182

45. Katie Bacon, "The Dark Side of the Gilded Age," *Atlantic*, June 2007, https://www .theatlantic.com/magazine/archive/2007/06/the-dark-side-of-the-gilded-age/306012/.

46. "Theodore Roosevelt Fourth Annual Message to Congress," December 6, 1904, https://millercenter.org/the-presidency/presidential-speeches/december-6 -1904-fourth-annual-message.

47. Tim Wu, *The Curse of Bigness*, pp. 24–25.

48. Sherman Antitrust Act, 15 U.S.C., July 2, 1890.

49. Clayton Antitrust Act, 15 U.S.C., June 5, 1914, as amended.

50. Theodore Roosevelt, "Remarks at the 42nd Anniversary Banquet of the Union League Club in Philadelphia," American Presidency Project, January 30, 1905, American Presidency Project, www.presidency.ucsb.edu/documents/remarks-the-forty-second -anniversary-banquet-the-union-league-club-philadelphia.

CHAPTER 2

1. Arnold Toynbee, *Lectures on the Industrial Revolution of the 18th Century in England* (London: Rivington's, 1887, repr. Cambridge University Press, 2011).

2. Ibid., p. 85.

3. Klaus Schwab, *The Fourth Industrial Revolution* (New York: Crowd Business, 2016).

4. Ibid., p. 4.

5. Ibid., p. 2.

6. Jason Fernando, "Factors of Production," *Investopedia*, July 30, 2021, https://www .investopedia.com/terms/f/factors-production.asp.

7. "What Tech Does China Want?," *Economist*, August 9, 2021, https://www.economist.com/busness/what-does-china-want/21803410.

8. Marco Iansiti and Karim L. Lakhani, "Competing in the Age of AI," *Harvard Business Review*, February 2020.

9. Ibid.

10. Anchal Gupta, "How Do Platform Businesses Win?," *Medium*, February 10, 2019, https://anchalgupt25.medium.com/why-do-platform-businesses-win-46485be8863b.

11. Jean Tirole, "Regulating the Disrupters," *ING: Think Outside*, January 4, 2019, https://think.ing.com/downloads/pdf/opinion/jean-tirole-regulating-the-disrupters.

12. Tom Wheeler, *From Gutenberg to Google: The History of Our Future* (Washington, DC: Brookings Institution Press, 2019), pp. 223–226.

13. Carl Benedikt Frey and Michael Osborne, "Technology at Work: The Future of Innovation and Employment," *Citi GPS: Global Perspectives & Solutions* (Oxford University), February 2015, p. 16.

14. Walter A. McDougall, *Throes of Democracy: The American Civil War Era 1829–1877* (New York: HarperCollins, 2009), p. 148.

15. Ibid.

16. See Wheeler, *From Gutenberg to Google*, pp. 72–74 and 194–199.

CHAPTER 3

1. For a discussion of Theodore Vail and the evolution of the Bell Telephone Company to AT&T, see Tom Wheeler, *From Gutenberg to Google: The History of Our Future* (Washington, DC: Brookings Institution Press, 2019), pp. 141-150.

2. Paul Baran, "On Distributed Communications," RAND Corporation, Memorandum RM-3420-PR, August 1964, https://www.rand.org/content/dam/rand/pubs/research_memoranda/2006/RM3420.pdf.

3. Stewart Brand, "Founding Father," *Wired*, March 1, 2001, https://www.wired.com/2001/03/baran/.

4. The 2015 Open Internet Order by the Obama Federal Communications Commission was repealed by the Trump FCC. As of this writing, the Biden FCC is attempting to reinstate the rule.

5. "Statement of Chair Lina M. Khan Regarding the Report to Congress on Privacy and Security," Federal Trade Commission, October 1, 2021, https://www.ftc.gov/public-statements/2021/10/statement-chair-lina-m-khan-regarding-report-congress-privacy-security.

6. Mason Walker and Katerina Eva Matsa, "News Consumption across Social Media in 2021," Pew Research Center, September 20, 2021, https://www.pewresearch.org/journalism/2021/09/20/news-consumption-across-social-media-in-2021/.

7. Casey Newton, "Mark in the Metaverse," *The Verge*, July 22, 2021, https://www.theverge.com/22588022/mark-zuckerberg-facebook-ceo-metaverse-interview.

8. Sheera Frenkel and Cecilia Kang, *An Ugly Truth: Inside Facebook's Battle for Domination* (New York: HarperColins, 2021), p. 285.

9. Kevin Roose, "Facebook Reverses Postelection Algorithm Changes That Boosted News from Authoritative Sources," *New York Times*, December 16, 2020, https://

www.nytimes.com/2020/12/16/technology/facebook-reverses-postelection-algorithm
-changes-that-boosted-news-from-authoritative-sources.html.

10. Frenkel and Kang, *An Ugly Truth*, p. 285.

11. Ibid., p. 286.

12. Craig Timberg, Elizabeth Dwoskin, and Reed Albergotti, "Inside Facebook, Jan. 6 Violence Fueled Anger, Regret over Missed Signs," *Washington Post*, October 23, 2021, https://www.washingtonpost.com/technology/2021/10/22/jan-6-capitol-riot-facebook/.

13. Ibid.

14. Oversight Board, Facebook, "Case Decision 2021–001-FB-FBR," May 5, 2021, https://oversightboard.com/decision/FB-691QAMHJ/.

15. "Facebook Responses to Oversight Board, Recommendation 14," https://about .fb.com/wp-content/uploads/2021/06/Facebook-Responses-to-Oversight-Board -Recommendations-in-Trump-Case.pdf.

16. Michelle Castillo, "Zuckerberg Tells Congress Facebook Is Not a Media Company: 'I consider us to be a technology company,'" CNBC, April 11, 2018, https:// www.cnbc.com/2018/04/11/mark-zuckerberg-facebook-is-a-technology-company-not -media-company.html.

CHAPTER 4

1. "Facebook to Acquire Oculus," Meta, March 25, 2014, https://about.fb.com/news /2014/03/facebook-to-acquire-oculus/. See also https://en.wikipedia.org/wiki/Oculus _(brand).

2. Facebook Second Quarter 2021 Results, Conference Call Transcript, July 28, 2021, https://s21.q4cdn.com/399680738/files/doc_financials/2021/q2/FB-Q2-2021-Earnings -Call-Transcript.pdf.

3. Salvador Rodriguez, "Facebook Changes Company Name to Meta," CNBC, October 28, 2021, https://www.cnbc.com/2021/10/28/facebook-changes-company-name-to -meta.html.

4. Neal Stephenson, *Snow Crash* (New York: Bantam Books, 1992).

5. Eric M. Johnson and Tim Hepher, "Boeing Wants to Build Its Next Airplane in the 'Metaverse,'" Reuters, December 17, 2021, https://www.reuters.com/technology/boeing -wants-build-its-next-airplane-metaverse-2021-12-17/.

6. Ina Fried, "For Meta, New Headset But Same Old Problems," *Axios*, October 12, 2022, https://www.axios.com/newsletters/axios-login-27612ce2-a550-4957-86d0 -04b98b4186a4.html?utm_source=newsletter&utm_medium=email&utm_campaign =newsletter_axioslogin&stream=top.

7. "Gartner Predicts 25% of People Will Spend At Least One Hour Per Day in the Metaverse by 2026," Gartner, February 7, 2022, https://www.gartner.com/en/newsroom /press-releases/2022-02-07-gartner-predicts-25-percent-of-people-will-spend-at-least -one-hour-per-day-in-the-metaverse-by-2026.

8. Jeff Horowitz, "Company Documents Show Meta's Flagship Metaverse Falling Short," *Wall Street Journal*, October 15, 2022, https://www.wsj.com/articles/meta -metaverse-horizon-worlds-zuckerberg-facebook-internal-documents-11665778961.

9. Mike Robuck, "Meta to Spend $19.2B in Metaverse Next Year," Mobile World Live, December 20, 2022, https://www.mobileworldlive.com/featured-content/home-banner/meta-to-invest-19-2b-in-metaverse-next-year/.

10. Discussion with author, November 16, 2022.

11. ACM Symposium on Eye Tracking Research & Applications, Seattle, June 8–11, 2022, https://etra.acm.org/2022/.

12. Jacob Leon Kroger, Otto Hans-Martin Lutz, and Florian Muller, "What Does Your Gaze Reveal about You? On the Privacy Implications of Eye Tracking," *IFIP Advances in Information and Communication Technology*, vol. 576, Springer, https://link.springer.com/chapter/10.1007/978-3-030-42504-3_15#citeas.

13. Philip Parnamets, Petter Johansson, Lars Hall, and Daniel C. Richardson, "Biasing Moral Decisions by Exploiting the Dynamics of Eye Gaze," *PNAS* 112, no. 13 (2015): 4170-4175, https://www.pnas.org/doi/full/10.1073/pnas.1415250112.

14. Hannah Murphy, "Facebook Patents Reveal How It Intends to Cash in on Metaverse," *Financial Times*, January 18, 2022, https://www.ft.com/content/76d40aac-034e-4e0b-95eb-c5d34146f647.

15. Louis Rosenberg, "Metaverse: Augmented Reality Pioneer Warns It Could Be Far Worse Than Social Media," Big Think, November 6, 2021, https://bigthink.com/the-future/metaverse-augmented-reality-danger/.

16. Jonathan Vanian, "Mark Zuckerberg's Metaverse May Be as Privacy Flawed as Facebook," *Fortune*, October 29, 2021, https://fortune.com/2021/10/29/mark-zuckerberg-metaverse-privacy-facebook-meta/.

17. Steve Kovach, "Here's How Zuckerberg Thinks Facebook Will Profit by Building a 'Metaverse,'" CNBC, July 29, 2021, https://www.cnbc.com/2021/07/29/facebook-metaverse-plans-to-make-money.html.

18. Nick Clegg, "Making the Metaverse: What It Is, How It Will Be Built, and Why It Matters," Meta, May 18, 2022, https://tech.facebook.com/ideas/2022/05/making-the-metaverse/.

19. Casey Newton, "Mark in the Metaverse," *The Verge*, July 22, 2021, https://www.theverge.com/22588022/mark-zuckerberg-facebook-ceo-metaverse-interview.

20. Waltzman quoted in Andrew S. Chow, "A Year Ago, Facebook Pivoted to the Metaverse. Was It Worth It?," *Time*, October 27, 2022, https://time.com/6225617/facebook-metaverse-anniversary-vr/.

21. Sheera Frenkel and Kellen Browning, "The Metaverse's Dark Side: Here Come Harassment and Assaults," *New York Times*, December 30, 2021, https://www.nytimes.com/2021/12/30/technology/metaverse-harassment-assaults.html.

22. "Facebook's Metaverse," Center for Countering Digital Hate, December 30, 2021, https://counterhate.com/research/facebooks-metaverse/?_ga=2.39938658.1985441126.1665761340-1467858480.1665761340&_gl=1%2A1vr8yuk%2A_ga%2AMTQ2Nzg1ODQ4MC4xNjY1NzYxMzQw%2A_ga_V7WR404SEC%2AMTY2NTc2MTQwOS4xLjEuMTY2NTc2MTQ0MC42MC4wLjA.

23. Yinka Bokinni, "A Barrage of Assault, Racism and Rape Jokes: My Nightmare Trip into the Metaverse," *Guardian*, April 25, 2022, https://www.theguardian.com/tv

-and-radio/2022/apr/25/a-barrage-of-assault-racism-and-jokes-my-nightmare-trip-into-the-metaverse.

24. Lauren Farrar, "Is the Internet Making You Meaner?," KQED, August 5, 2019, https://www.kqed.org/education/532334/is-the-internet-making-you-meaner#:~:text=Yes.,in%20face%20to%20face%20conversations.

25. Morning Consult, Twitter, April 16, 2022, https://twitter.com/MorningConsult/status/1515451970781392906.

26. Randeep Sudan, "How Should Governments Prepare for the Metaverse?," Digital Diplomacy, July 20, 2021, https://medium.com/digital-diplomacy/how-should-governments-prepare-for-the-metaverse-90fd03387a2a.

27. Litska Strikwerda, "Theft of Virtual Items in Online Multiplayer Computer Games: An Ontological and Moral Analysis," *Ethics and Information Technology*, January 6, 2012, https://link.springer.com/article/10.1007/s10676-011-9285-3.

28. Angus Crawford and Bethan Bell, "Molly Russell Inquest: Father Makes Social Media Plea," *BBC News*, September 30, 2022, https://www.bbc.com/news/uk-england-london-63073489.

29. Terence Shin, "Real-Life Examples of Discriminating Artificial Intelligence," Towards Data Science, June 4, 2020, https://towardsdatascience.com/real-life-examples-of-discriminating-artificial-intelligence-cae395a90070.

30. Nicol Turner Lee, "It's Time for an Updated Big Tech Civil Rights Regime," PolicyLink, https://www.policylink.org/sites/default/files/Nicol_Turner_Lee_082522a.pdf.

31. "Racial Discrimination in the Sharing Economy," HBS Working Knowledge, *Forbes*, February 24, 2014, https://www.forbes.com/sites/hbsworkingknowledge/2014/02/24/racial-discrimination-in-the-sharing-economy/?sh=1d85d64b5c44.

32. Jillian Deutsch, Naomi Nix, and Sara Kopit, "Misinformaton Has Already Made Its Way to the Metaverse," Bloomberg, December 15, 2021, https://www.bloomberg.com/press-releases/2021-12-15/misinformaton-has-already-made-its-way-to-facebook-s-metaverse.

CHAPTER 5

1. OpenAI, "GPT-4," March 14, 2023, https://openai.com/research/gpt-4.

2. Zuckerberg quoted in Catherine Thorbecke, "What Metaverse? Meta Says Its Single Largest Investment Is Now in 'Advancing AI,'" CNN, March 15, 2023, https://www.cnn.com/2023/03/15/tech/meta-ai-investment-priority/index.html.

3. "Artificial Intelligence Could Be Our Savior, According to the CEO of Google," World Economic Forum, January 24, 2018. See the website at https://www.weforum.org/agenda/2018/01/google-ceo-ai-will-be-bigger-than-electricity-or-fire/.

4. See the website at https://twitter.com/StevenLevy/status/1631442434168168448?ref_src=twsrc%5Egoogle%7Ctwcamp%5Eserp%7Ctwgr%5Etweet.

5. "2022 Expert Survey on Progress in AI," *AI Impacts*, August 3, 2022, https://aiimpacts.org/2022-expert-survey-on-progress-in-ai/#Extinction_from_AI.

6. Pichai quoted in "Artificial Intelligence Will Save Us, Not Destroy Us; Google CEO," NewsBharati, February 7, 2018, https://www.newsbharati.com/Encyc/2018/2/7/Google-on-AI.html.

7. Jacob Stern, "Five Remarkable Chats That Will Help You Understand ChatGPT," *The Atlantic*, December 8, 2022, https://www.theatlantic.com/technology/archive/2022 /12/openai-chatgpt-chatbot-messages/672411/.

8. Thomas H. Ptacek, Twitter, December 1, 2022, https://twitter.com/tqbf/ status/1598513757805858820.

9. Jonathan Vanian, "Microsoft Adds OpenAI Technology to Word and Excel," CNBC, March 16, 2023, https://www.cnbc.com/2023/03/16/microsoft-to-improve -office-365-with-chatgpt-like-generative-ai-tech-.html.

10. Sundar Pichai, "An Important Next Step on Our AI Journey," Google, February 6, 2023, https://blog.google/technology/ai/bard-google-ai-search-updates/ and Jeremy Khan, "GPT-4 Debuts and Google Beats Microsoft in a Race to Add Generative A.I. to Consumer Office Tools," *Fortune*, March 14, 2023, https://fortune.com/2023/03/14 /gpt-4-debuts-and-google-beats-microsoft-in-race-to-add-generative-a-i-to-consumer -office-tools/.

11. Ben Wodeski, "Meta: LLaMA Language Model Outperforms OpenAI's GPT-3," AI Business, February 27, 2023, https://aibusiness.com/meta/meta-s-llama-language -model-outperforms-openai-s-gpt-3.

12. See the website at https://www.anthropic.com/.

13. See the website at https://stablediffusionweb.com/.

14. Adam Lashinsky, "Silicon Valley Faces Another Make-or-Break Moment," *Washington Post*, March 20, 2023, https://www.washingtonpost.com/opinions/2023/03/20/ chatgpt-ai-silicon-valley-winners/.

15. Andrew R. Chow, "How ChatGPT Managed to Grow Faster Than Instagram or TikTok," *Time*, February 8, 2023, https://time.com/6253615/chatgpt-fastest-growing/.

16. "How Much Time It Took to Reach 100 Million Users," *The Hacking News* (blog), July 28, 2015, https://latesthackingnews.com/2015/07/28/how-much-time-it-took-to -reach-100-million-users/.

17. Dina Bass, "Microsoft Strung Together Tens of Thousands of Chips in a Pricey Supercomputer for OpenAI," Bloomberg, March 13, 2023, https://www.bloomberg .com/news/articles/2023-03-13/microsoft-built-an-expensive-supercomputer-to-power -openai-s-chatgpt?sref=ExbtjcSG.

18. Ibid.

19. OpenAI, "GPT-4 Technical Report," March 16, 2023, p. 2, https://cdn.openai.com /papers/gpt-4.pdf.

20. Catherine Shu, "Google Acquires Artificial Intelligence Startup DeepMind for More Than $500M," TechCrunch, January 26, 2014, https://techcrunch.com/2014/01 /26/google-deepmind/.

21. James Somers, "How the Artificial Intelligence Program AlphaGo Mastered Its Games," *New Yorker*, December 28, 2018, https://www.newyorker.com/science/elements/ how-the-artificial-intelligence-program-alphazero-mastered-its-games.

22. Google, "Our Principles," https://ai.google/principles/.

23. Ibid.

24. See the website at https://twitter.com/paultoo/status/1598434161332981760?lang =en.

25. Ben Cost, "Rise of the Bots: 'Scary' AI ChatGPT Could Eliminate Google within Two Years," *New York Post*, December 6, 2022, https://nypost.com/2022/12/06/scary -chatgpt-could-render-google-obsolete-in-two-years/.

26. Miles Kruppa and Sam Schechner, "How Google Became Cautious of AI and Gave Microsoft an Opening," *Wall Street Journal*, March 7, 2023, https://www.wsj.com/ articles/google-ai-chatbot-bard-chatgpt-rival-bing-a4c2d2ad.

27. Ibid.

28. Victor Ordoñez, Taylor Dunn, and Eric Noll, "OpenAI CEO Sam Altman Says AI Will Reshape Society, Acknowledges Risks: 'A Little Bit Scared of This,'" *ABC News*, March 16, 2023, https://abcnews.go.com/Technology/openai-ceo-sam-altman-ai -reshape-society-acknowledges/story?id=97897122.

29. "U.S. Chamber of Commerce Calls for AI regulation," Reuters, March 9, 2023, https://www.reuters.com/technology/us-chamber-commerce-calls-ai-regulation-2023 -03-09/

30. See website at https://www.whitehouse.gov/ostp/ai-bill-of-rights/.

31. See website at https://eur-lex.europa.eu/legal-content/EN/TXT/?uri=CELEX: 52021PC0206.

32. European Commission, "Regulatory Framework Proposal on Artificial Intelligence," accessed March 24, 2023, https://digital-strategy.ec.europa.eu/en/policies/ regulatory-framework-ai.

33. Zheping Huang, "ChatGPT Lookalikes Proliferate in China," Bloomberg, February 15, 2023, https://www.bloomberg.com/news/newsletters/2023-02-15/chatgpt-is -china-s-new-obsession-as-lookalikes-proliferate-on-wechat.

34. "China Prepares to Police AI as ChatGPT Frenzy Spreads," Bloomberg, February 24, 2023, https://www.bloomberg.com/news/articles/2023-02-24/china-prepares-to -police-ai-as-chatgpt-frenzy-spreads.

35. Ibid.

36. OpenAI, "GPT-4 Technical Report," p. 14.

37. Ibid., p. 2.

38. Karen Hao, "AI Is Sending People to Jail—and Getting It Wrong," *MIT Technology Review*, January 21, 2019, https://www.technologyreview.com/2019/01/21/137783/ algorithms-criminal-justice-ai/.

39. Stuart A. Thompson, Tiffany Hsu, and Steven Lee Myers, "Conservatives Aim to Build a Chatbot of Their Own," *New York Times*, March 22, 2023, https://www.nytimes .com/2023/03/22/business/media/ai-chatbots-right-wing-conservative.html.

40. Jessica Guynn, "Why Elon Musk Wants to Build ChatGPT Competitor: AI Chatbots Are Too 'Woke,'" *USA Today*, March 1, 2023, https://www.usatoday.com/story/tech /2023/03/01/woke-chatgpt-ai-musk-chatbot/11375469002/.

41. Geoffrey Fowler, "Trying Microsoft's New AI Chatbot Search Engine, Some Answers Are Uh-Oh," *Washington Post*, February 8, 2023, https://www.washingtonpost .com/technology/2023/02/07/microsoft-bing-chatgpt/.

42. Reneé DiResta, "The Supply of Disinformation Will Soon Be Infinite," *The Atlantic*, September 20, 2020, https://www.theatlantic.com/ideas/archive/2020/09/future -propaganda-will-be-computer-generated/616400/.

43. Alex Hughes, "ChatGPT: Everything You Need to Know about OpenAI's GPT-4 Tool," *BBC Science Focus*, March 16, 2023, https://www.sciencefocus.com/future-technology/gpt-3/.

44. OpenAI, "Privacy Policy," March 14, https://openai.com/policies/privacy-policy.

45. Ibid.

46. Ibid.

47. OpenAI, "Usage Policies," March 23, 2023, https://openai.com/policies/usage-policies.

48. Bernard Marr, "How Dangerous Are ChatGPT and Natural Language Technology for Cybersecurity?," *Forbes*, January 25, 2023, https://www.forbes.com/sites/bernardmarr/2023/01/25/how-dangerous-are-chatgpt-and-natural-language-technology-for-cybersecurity/?sh=1f01076c4aa6.

49. Alex Hern, "TechScape: Will Meta's Massive Leak Democratise AI—and at What Cost?," *Guardian*, March 7, 2023, https://www.theguardian.com/technology/2023/mar/07/techscape-meta-leak-llama-chatgpt-ai-crossroads.

50. Victor Ordoñez, Taylor Dunn, and Eric Noll, "OpenAI CEO Sam Altman Says AI Will Reshape Society, Acknowledges Risks: 'A Little Bit Scared of This,'" *ABC News*, March 16, 2023.

51. Steven Levy, "How to Start an AI Panic," *Wired*, March 18, 2023, https://www.wired.com/story/plaintext-how-to-start-an-ai-panic/.

WHO MAKES THE RULES?

1. Cicero, "Omnes Legum Servi Sumus," *Right Reason*, https://rightreason.typepad.com/right_reason/m-tullius-cicero.

CHAPTER 6

1. Ultimately, Zuckerberg realized its arrogance and the Facebook motto was changed. The attitude it represents, however, lives on.

2. Janko Roettgers, "Twitter CEO Admits Company Didn't Fully Grasp Abuse Problem," *Variety*, March 1, 2018, https://variety.com/2018/digital/news/twitter-ceo-abuse-1202714236/.

3. LoveMovieQuotes, *Jurassic Park*, "Your scientists were so preoccupied" (scene), YouTube video, https://www.youtube.com/watch?v=g3j9muCo4o0.

4. "How Connecting 7 Billion to the Web Will Transform the World," *PBS News Hour*, May 2, 2013, https://www.pbs.org/newshour/science/in-new-digital-age-google-leaders-see-more-possibilities-to-connect-the-worlds-7-billion; Eric Schmidt and Jared Cohen, *The New Digital Age: Transforming Nations, Businesses, and Our Lives* (Knopf Doubleday, 2013), p. 3.

5. John Steele Gordon, *An Empire of Wealth: The Epic History of American Economic Power* (New York: Harper Perennial, 2004), p. 236.

6. Susan Berfield, *The Hour of Fate: Theodore Roosevelt, J.P. Morgan, and the Battle to Transform American Capitalism* (New York: Bloomsbury, 2020), p. 46.

7. Ibid., p. 46.

8. Doris Kearns Goodwin, *The Bully Pulpit: Theodore Roosevelt, William Howard Taft, and the Golden Age of Journalism* (New York: Simon & Schuster, 2013), p. 297.

9. Gordon, *An Empire of Wealth*, p. 262.

10. Berfield, *The Hour of Fate*, p. 126.

11. Ben Yagoda, "A Short History of 'Hack,'" *New Yorker*, March 6, 2014, https://www.newyorker.com/tech/annals-of-technology/a-short-history-of-hack.

12. The EU and the state of California have adopted online privacy protection rules. These have a de facto rather than a de jure impact on the behavior of companies outside their jurisdictions. The Federal Trade Commission (FTC) in 2022 initiated a process that could lead to new data privacy protections, but its procedures and the court review that will follow any action mean final resolution is years away. As of this writing, the U.S. Congress has been unable to pass comprehensive digital privacy legislation.

13. Bobbie Johnson, "Privacy No Longer a Social Norm, Says Facebook Founder," *Guardian*, January 10, 2010, https://www.theguardian.com/technology/2010/jan/11/facebook-privacy.

14. "Google Receives $25 Million in Equity Funding," *News from Google*, June 7, 1999.

15. Ibid.

16. Derek Thompson, "Google's CEO: 'The Laws Are Written by Lobbyists,'" *Atlantic*, October 1, 2010, https://www.theatlantic.com/technology/archive/2010/10/googles-ceo-the-laws-are-written-by-lobbyists/63908/.

17. Ibid.

18. Alexis Madrigal, "Reading the Privacy Policies You Encounter in a Year Would Take 76 Work Days," *Atlantic*, March 1, 2012, https://www.theatlantic.com/technology/archive/2012/03/reading-the-privacy-policies-you-encounter-in-a-year-would-take-76-work-days/253851/.

19. Geoffrey A. Fowler, "No, Mark Zuckerberg, We're Not Really in Control of Our Data," *Washington Post*, April 12, 2018, https://washingtonpost.com/news/the-switch/wp/2018/04/12/no-mark-zuckerberg-were-not-really-in-control-of-our-data/.

20. Ryan Nakashima, "AP Exclusive: Google Tracks Your Movements, Like It or Not," *AP News*, August 13, 2018, https://apnews.com/828aefab64d4411bac257a07c1af0ecb.

21. Yves-Alexandre de Montjoye et al., "Unique in the Crowd: The Privacy Bounds of Human Mobility," *Scientific Reports* (March 2013), art. 1376, https://www.nature.com/articles/srep01376.

22. Edmond Locard, *Traité de Criminalistique* (Lyon: J. Desvigne et ses fils, 1931).

23. Douglas C. Schmidt, *Google Data Collection*, Digital Content Next, August 15, 2018, https://digitalcontentnext.org/blog/2018/08/21/google-data-collection-research/.

24. Casey Newton, "Tim Cook: Silicon Valley's Most Successful Companies Are Selling You Out," *The Verge*, June 2, 2015, https://www.theverge.com/2015/6/2/8714345/tim-cook-epic-award-privacy-security.

25. James Vincent, "Tim Cook Warns of 'Data-Industrial Complex' in Call for Comprehensive US Privacy Laws," *The Verge*, October 24, 2018, https://www.theverge.com/2018/10/24/18017842/tim-cook-data-privacy-laws-us-speech-brussels.

26. Pew Research Center, "The State of Privacy in Post-Snowden America," September 21, 2016, https://www.pewresearch.org/fact-tank/2016/09/21/the-state-of-privacy-in -america/.

27. Roger McNamee, *Zucked: Waking Up to the Facebook Catastrophe* (New York: Penguin Press, 2019), p. 95.

28. See Meta Connectivity, "Free Basics," https://www.facebook.com/connectivity/ solutions/free-basics.

29. Toussaint Nothias, "Access Granted: Facebook's Free Basics in Africa," *Media, Culture & Society* (April 2020), https://journals.sagepub.com/doi/full/10.1177 /0163443719890530.

30. Author discussion with Nikhil Pahwa, Indian journalist and digital rights activist, Harvard Kennedy School, February 9, 2018.

31. Vindu Goel and Mike Isaac, "Facebook Loses a Battle in India over Its Free Basics Program," *New York Times*, February 8, 2016, https://www.nytimes.com/2016/02/09/ business/facebook-loses-a-battle-in-india-over-its-free-basics-program.html.

32. Goodwin, *The Bully Pulpit*, p. 448.

33. Ibid.

34. Ibid.

35. Sean Burch, "'Senator, We Run Ads': Hatch Mocked for Basic Facebook Question to Zuckerberg," *The Wrap*, April 10, 2018, https://www.thewrap.com/senator-orrin-hatch -facebook-biz-model-zuckerberg/.

36. Goodwin, *The Bully Pulpit*, p. 456, citing the papers of Ray Baker, the legendary muckraker who wrote a six-part exposé, "The Railroads on Trial," for *McClure's Magazine*.

37. John D. McKinnon and Chad Day, "Tech Companies Make Final Push to Head Off Tougher Regulation," *Wall Street Journal*, December 19, 2002, https://www.wsj.com/articles /tech-companies-make-final-push-to-head-off-tougher-regulation-11671401283.

38. Computer and Communications Industry Association, https://dontbreakwhatworks .ccianet.org/

39. Consumer Technology Association, *Don't Undermine America's National Security*, YouTube video, https://www.youtube.com/watch?v=dmldyGqD2V4.

40. Goodwin, *The Bully Pulpit*, p. 455.

41. Mark Scott, "Google and Apple as Privacy Regulators 2.0," *Politico Digital Bridge*, June 10, 2021, https://www.politico.eu/newsletter/digital-bridge/politico-digital-bridge -eu-us-trade-and-tech-council-g7-explainer-google-apple-privacy-2-0/.

42. Greg Bensinger, "Google's Privacy Backpedal Shows Why It's So Hard Not to Be Evil," *New York Times*, June 14, 2021, https://www.nytimes.com/2021/06/14/opinion/ google-privacy-big-tech.html?referringSource=articleShare.

43. Mike Isaac and Jack Nicas, "Breaking Point: How Mark Zuckerberg and Tim Cook Became Foes," *New York Times*, April 26, 2021, https://www.nytimes.com/2021/04 /26/technology/mark-zuckerberg-tim-cook-facebook-apple.html.

44. Quoted in Jack Beatty, *Colossus: How the Corporation Changed America* (New York: Broadway Books, 2001), pp. 132–33.

45. Goodwin, *The Bully Pulpit*, p. 299.

46. Ibid.

47. For the origin of this expression, see Quote Investigator, "Your Liberty to Swing Your Fist Ends Just Where My Nose Begins," https://quoteinvestigator.com/2011/10/15 /liberty-fist-nose/.

CHAPTER 7

1. "What Goes Around," *Economist*, Special Report, September 15, 2016, https://www .economist.com/special-report/2016/09/15/what-goes-around.

2. Ibid.

3. Statista, "Share of Amazon, Facebook, and Google in Net Digital Ad Revenue in the United States from 2019 to 2023," https://www.statista.com/statistics/242549/digital-ad -market-share-of-major-ad-selling-companies-in-the-us-by-revenue.

4. Nicole Farley, "Search Advertising Is Thriving amid Economic Uncertainty, New Report Confirms," Search Engine Land, September 16, 2022.

5. Statista, "Android Operating System Share Worldwide by OS Version from 2013–2017," https://www.statista.com/statistics/271774/share-of-android-platforms-on -mobile-devices-with-android-os/.

6. "Amazon's Share of US eCommerce Sales Hits All-Time High of 56.7% in 2021," March 14, 2022, PYMNTS, https://www.pymnts.com/news/retail/2022/amazons-share -of-us-ecommerce-sales-hits-all-time-high-of-56-7-in-2021/.

7. Rahul Kumar, *WPOven*, "AWS Market Share 2022: How Far It Rules the Could Industry?," September 20, 2022, https://www.wpoven.com/blog/aws-market-share/.

8. Joe Lonsdale, "The Case for Splitting Amazon in Two," *Wall Street Journal*, February 7, 2002, https://www.wsj.com/articles/split-amazon-in-two-prime-web-services-aws -logistics-third-party-earnings-report-consumers-antitrust-11644249482.

9. Mix Dimitar, "Facebook Owns Four out of the Five Most Downloaded Apps Worldwide," Next Web, April 18, 2017, https://thenextweb.com/apps/2017/04/18/ facebook-downloaded-app-netflix/.

10. Peter Roesler, "90% of Website Social Logins Are with Facebook or Google," Web Marketing Pros, https://www.webmarketingpros.com/90-of-website-social-logins-are -with-facebook-or-google/.

11. Joel Hruska, "Microsoft Bleeds Market Share as Chromebooks Outsell Macs for the First Time," Extreme Tech, February 18, 2021, https://www.extremetech.com /computing/320088-microsoft-bleeds-market-share-as-chromebooks-outsell-macs-for -the-first-time.

12. Shira Ovide, "What Big Tech's Riches Mean for Our Future," *New York Times*, February 3, 2022, https://www.nytimes.com/2022/02/03/technology/big-tech-facebook -earnings.html?referringSource=articleShare.

13. Scott Rosenberg, "Tech Giants Are the New Gatekeepers," *Axios*, February 1, 2019, www.axios.com/tech-giants-new-gatekeepers-1548976974-25f26494-a67c-4252-9c18 -418588f8de06.html.

14. Nariman Haghighi, "Scott Galloway's Powerful Podcast on Amazon, Future of Retail, E-commerce & Tech Disruption," *CandidiO* (blog), Get Candid, June 18, 2017, https://www.getcandid.com/blog/2017/06/18/scott-galloways-powerful-podcast -on-amazon-future-of-retail-e-commerce-tech-disruption/.

15. "Digital Transformation Is Racing Ahead and No Industry Is Immune," *Harvard Business Review*, July 19, 2017, https://hbr.org/sponsored/2017/07/digital-transformation-is-racing-ahead-and-no-industry-is-immune-2.

16. "Global 500," *Financial Times*, May 2001, https://web.archive.org/web/20071113124026/http:/specials.ft.com/ft500/may2001/FT36H8Z8KMC.html.

17. Statista, "The 100 Largest Companies in the World by Market Capitalization in 2021," 2021, https://www.statista.com/statistics/263264/top-companies-in-the-world-by-market-capitalization/.

18. Douglas Edwards, *I'm Feeling Lucky: The Confessions of Google Employee Number 59* (New York: Houghton Mifflin Harcourt, 2011), p. 190.

19. Louise Story and Miguel Helft, "Google Buys DoubleClick for $3.1 Billion," *New York Times*, April 14, 2007, https://www.nytimes.com/2007/04/14/technology/14DoubleClick.html.

20. John Callaham, "Google Made Its Best Acquisition Nearly 16 Years Ago: Can You Guess What It Was?," Android Authority, May 21, 2021, https://www.androidauthority.com/google-android-acquisition-884194/.

21. Serving an unserved market where access to existing data is not essential is exemplified by TikTok, which found a way around the data dominance of others with a new product that developed its own data by dangling the promise of celebrity to users.

22. TikTok, for instance, was able to grow rapidly by identifying a new market—young people—where beginning the business did not have to rely on the data Facebook or YouTube was using for its services. The use of the TikTok app then began the creation of a new data hoard accompanied by national security concerns as to the safety of data held by the Chinese-owned company.

23. Alvin Toffler and Heidi Toffler, *Revolutionary Wealth* (New York: Alfred A. Knopf, 2006), p. 168.

24. Shona Ghosh, "Sheryl Sandberg Just Dodged a Question about Whether Facebook Is a Media Company," *Business Insider*, October 12, 2017, https://markets.businessinsider.com/news/stocks/sheryl-sandberg-dodged-question-on-whether-facebook-is-a-media-company-2017-10.

25. Elisa Shearer and Amy Mitchell, "News Use Across Social Media Platforms in 2020," Pew Research Center, January 12, 2021, https://www.pewresearch.org/journalism/2021/01/12/news-use-across-social-media-platforms-in-2020/.

26. Roger McNamee, "How to Fix Facebook: Make Users Pay for It," *Washington Post*, February 21, 2018.

27. Peter Thiel, "Competition Is for Losers," *Wall Street Journal*, September 12, 2014, https://www.wsj.com/articles/peter-thiel-competition-is-for-losers-1410535536.

28. Natalie Sherman, "Facebook Facing US Legal Action over Competition," *BBC News*, December 9, 2020, https://www.bbc.com/news/business-55250366.

29. Om Malik, "Here Is Why Facebook Bought Instagram," GigaOm, April 9, 2012, https://gigaom.com/2012/04/09/here-is-why-did-facebook-bought-instagram/.

30. Ingrid Lunden, "Facebook Buys Mobile Data Analytics Company Onavo Reportedly for Up to $200 M . . . and (Finally?) Gets Its Office in Israel," TechCrunch, October 14, 2013, https://techcrunch.com/2013/10/13/facebook-buys-mobile-analytics-company-onavo-and-finally-gets-its-office-in-israel/.

31. Charlie Warzel and Ryan Mac, "These Confidential Charts Show Why Facebook Bought WhatsApp," *BuzzFeed News*, December 5, 2018, https://www.buzzfeednews.com /article/charliewarzel/why-facebook-bought-whatsapp.

32. Ibid.

33. Carly Hallman, "Everything Facebook Owns: Mergers and Acquisitions from the Past 15 Years," TitleMax, https://www.titlemax.com/discovery-center/lifestyle/ everything-facebook-owns-mergers-and-acquisitions-from-the-past-15-years/.

34. Leena Rao, "Eric Schmidt: Google Is Buying One Company a Week," Tech-Crunch, December 7, 2011, https://techcrunch.com/2011/12/07/eric-schmidt-google-is -buying-one-company-a-week/.

35. Timothy B. Lee, "Emails Detail Amazon's Plan to Crush a Startup Rival with Price Cuts," Ars Technica, July 30, 2020, https://arstechnica.com/tech-policy/2020/07/emails -detail-amazons-plan-to-crush-a-startup-rival-with-price-cuts/.

36. Will Oremus, "A Classic Silicon Valley Tactic—Losing Money to Crush Rivals— Comes in for Scrutiny," *Washington Post*, July 6, 2021.

37. Dave Lavinsky, "$500 Million: Amazon Acquires Quidsi: Sell or Work for 341 Years?," *Growthink*, www.growthink.com/content/500-million-amazon-acquires-quidsi -sell-or-work-341-years; "Confirmed: Amazon Spends $545 Million On Diapers.com Parent Quidsi," TechCrunch, November 8, 2010, https://techcrunch.com/2010/11/08/ confirmed-amazon-spends-545-million-on-diapers-com-parent-quidsi/.

38. Oremus, "A Classic Silicon Valley Tactic."

39. Samuel T. Pees, "Oil History," Petroleum History Institute, 2004, http://www .petroleumhistory.org/OilHistory/pages/Whale/prices.html.

40. Schultzy, "Standard Oil—A Company So Effective, Only the U.S. Government Could Compete with It," paper, Harvard Business School, Harvard Business School sub-mission, December 2, 2015, https://digital.hbs.edu/platform-rctom/submission/standard -oil-a-company-so-effective-only-the-u-s-government-could-compete-with-it/.

41. Jack Beatty, *Colossus: How the Corporation Changed America* (New York: Broadway Books, 2001), p. 148.

42. Klobuchar, *Antitrust*, p. 101.

CHAPTER 8

1. See Daniel R. Headrick, *The Invisible Weapon: Telecommunications and International Politics, 1851–1945* (Oxford: Oxford University Press, 1991).

2. Wheeler, *From Gutenberg to Google*, p. 204.

3. Body of European Regulators for Electronic Communications (BEREC), "Guide-lines to National Regulatory Authorities (NRAs) on the implementation of the new net neutrality," August 30, 2016, https://berec.europa.eu/eng/news_and_publications/whats _new/3958-berec-launches-net-neutrality-guidelines.

4. The FTC does, however, possess the power to enforce interpretations of previously adopted statutes. Thus the Trump FTC levied the largest fine in the agency's history on Facebook because the company was deceiving its customers by not following its own privacy policies. Such enforcement is good; however, it was specific to Facebook and did

not deal with broader consumer privacy issues or impose any oversight on other platform companies.

5. Simon Van Dorpe and Leah Nylen, "Europe Failed to Tame Google. Can the U.S. Do Any Better?," *Politico*, October 21, 2020, https://www.politico.com/news/2020/10/21/google-europe-us-antitrust-431036.

6. California Consumer Privacy Act, https://oag.ca.gov/privacy/ccpa.

7. Daniel Van Boom, "China's Great Big Tech Experiment Matters Everywhere," CNET, November 3, 2021, https://www.cnet.com/news/chinas-great-big-tech-experiment/.

8. Geoffrey A. Manne, "A Comparative Look at Competition Law Approaches to Monopoly and Abuse of Dominance in the US and EU," United States Senate Committee on the Judiciary Subcommittee on Antitrust, Competition Policy, and Consumer Rights, December 19, 2018, https://www.judiciary.senate.gov/imo/media/doc/Manne Testimony.pdf.

9. European Court of Auditors, "The Commission's EU Merger Control and Antitrust Procedures: A Need to Scale Up Market Oversight," Special Report 24/2020, https://op.europa.eu/webpub/eca/special-reports/eu-competition-24-2020/en/.

10. Kelvin Chan, "After Years Grappling with Google, Europe Has Tips for US," Associated Press, October 21, 2020, https://apnews.com/article/google-antitrust-lawsuit-europe-tips-9b100e96d23849b742d27c457157b6bc.

11. Margrethe Vestager, "Competition in a Digital Age: Changing Enforcement for Changing Times," European Commission, June 26, 2020, https://ec.europa.eu/commission/commissioners/2019-2024/vestager/announcements/competition-digital-age-changing-enforcement-changing-times_en.

12. European Commission, "The Digital Markets Act: Ensuring Fair and Open Digital Markets," https://ec.europa.eu/info/strategy/priorities-2019-2024/europe-fit-digital-age/digital-markets-act-ensuring-fair-and-open-digital-markets_en.

13. See Tom Wheeler and Phil Verveer, "A Turning Point in the Oversight of Digital Platforms: A Challenge for American Leadership," Harvard Kennedy School, Shorenstein Center on Media, Politics, and Public Policy, February 25, 2021, https://shorensteincenter.org/turning-point-oversight-digital-platforms/.

14. European Comission, Digital Services Act, https://digital-strategy.ec.europa.eu/en/policies/digital-services-act-package.

15. Noah Feldman, "Free Speech in Europe Isn't What Americans Think," Bloomberg, March 19, 2017, https://www.bloomberg.com/opinion/articles/2017-03-19/free-speech-in-europe-isn-t-what-americans-think.

16. Government of the United Kingdom, *Unloacking Digital Competition: Report of the Digital Competition Expert Panel*, March 2019, https://assets.publishing.service.gov.uk/government/uploads/system/uploads/attachment_data/file/785547/unlocking_digital_competition_furman_review_web.pdf.

17. Dan Primack, "China Blocks Merger of Gaming Giants in Latest Tech Crackdown," *Axios*, July 12, 2021, https://www.axios.com/china-tech-crackdown-huya-douyu-396a9918-3555-46ec-a71f-d11eb9bf4baf.html.

18. Raymond Zhong, "China Fines Alibaba $2.8 Billion in Landmark Antitrust Case," *New York Times*, April 9, 2021, https://www.nytimes.com/2021/04/09/technology/china-alibaba-monopoly-fine.html.

19. Iris Deng, "Big Tech's 'Walled Gardens' Start to Crack as Tencent Vows to Follow Beijing's Order to Unblock Links to Rivals," *South China Morning Post*, September 13, 2021, https://www.scmp.com/tech/big-tech/article/3148572/big-techs-walled-gardens -start-crack-tencent-vows-follow-beijings.

20. Srishti Jha, "China Orders Tech Giants Alibaba and Tencent to Open Platforms Up to Each Other," Republic World, September 12, 2021, https://www.republicworld.com/business-news/international-business/china-orders-tech-giants-alibaba-and-tencent-to -open-platforms-up-to-each-other-reports.html.

21. Tom Wheeler, "China's New Regulation of Platforms: A Message for American Policymakers," *TechTank* (blog), Brookings Institution, September 14, 2021, https:// www.brookings.edu/blog/techtank/2021/09/14/chinas-new-regulation-of-platforms-a -message-for-american-policymakers/.

22. Meta Platforms, Inc., 10-K, https://investor.fb.com/financials/default.aspx.

Reasserting the Public Interest

1. Craig Timberg and Tony Room, "Facebook CEO Mark Zuckerberg to Capitol Hill: 'It Was My Mistake, and I'm Sorry,'" *Washington Post*, April 9, 2018, https://www .washingtonpost.com/news/the-switch/wp/2018/04/09/facebook-chief-executive-mark -zuckerberg-to-captiol-hill-it-was-my-mistake-and-im-sorry/.

Chapter 9

1. Gillian Tett, "The Human Factor—Why Data Is Not Enough to Understand the World," *Financial Times*, May 28, 2021, https://www.ft.com/content/4f00469c-75da -4e29-baf3-b7bec470732c.

2. Craig Timberg and Tony Romm, "Facebook CEO Mark Zuckerberg to Capitol Hill: 'It Was My Mistake, and I'm Sorry,'" *Washington Post*, April 9, 2018, https://www .washingtonpost.com/news/the-switch/wp/2018/04/09/facebook-chief-executive-mark -zuckerberg-to-captiol-hill-it-was-my-mistake-and-im-sorry/.

3. Tom Wheeler, Phil Verveer, and Gene Kimmelman, "New Digital Realities: New Oversight Solutions in the U.S.," discussion paper, Shorenstein Center, Harvard Kennedy School, Cambridge, MA, August 20, 2020, https://shorensteincenter.org/new-digital -realities-tom-wheeler-phil-verveer-gene-kimmelman/.

4. "Arthur Cecil Pigou," Wikipedia, https://en.wikipedia.org/wiki/Arthur_Cecil _Pigou.

5. Pigou's Cambridge professor, Alfred Marshall, was the first to develop the idea of externalities. Pigou then focused on the topic in his 1920 book, *The Economics of Welfare*.

6. Arthur Pigou, *The Economics of Welfare* (London: Macmillan, 1920; repr. New York: Routledge, 2017).

7. Julia Kagan, "Pigovian Tax," *Investopedia*, April 28, 2020, https://www.investopedia .com/terms/p/pigoviantax.asp.

8. Susan Berfield, *The Hour of Fate: Theodore Roosevelt, J. P. Morgan, and the Battle to Transform Amerian Capitalism* (New York: Bloomsbury, 2020), p. 48.

9. Tom Wheeler, *From Gutenberg to Google: The History of Our Future* (Washington, DC: Brookings Institution Press, 2019), p. 75.

10. Interstate Commerce Act of 1887, Public Law 49–41, February 4, 1887, https://www.ourdocuments.gov/print_friendly.php?flash=false&page=transcript&doc=49&title=Transcript+of+Interstate+Commerce+Act+%281887%29.

11. Communications Act of 1934, Public Law 73–416, June 19, 1934, https://transition.fcc.gov/Reports/1934new.pdf.

12. Federal Communications Commission, "FCC Releases Open Internet Order," 2015, https://www.fcc.gov/document/fcc-releases-open-internet-order.

13. In discussion with author at Aspen Institute Diplomacy and Technology seminar, August 2, 2017.

14. Of course, there are tangible results as well. However, the great value of these companies lies in their nonphysical assets. See Jonathan Haskel and Stian Westlake, *Capitalism without Capital: The Rise of the Intangible Economy* (Princeton: Princeton University Press, 2017).

15. Theodore Roosevelt, "Message Communicated to the Two Houses of Congress at the Beginning of the First Session of the Fifty-Seventh Congress," December 3, 1901, in *Addresses and Presidential Messages of Theodore Roosevelt, 1902–1904* (New York: G. P. Putnam's Sons, 1904), pp. 292–98, https://college.cengage.com/history/ayers_primary_sources/roosevelt_trusts_1901.htm.

16. Frederick Winslow Taylor, *The Principles of Scientific Management* (New York: Harper and Bros., 1911; repr. River Denys: NS: Ocean Minds Media House, 2020), pp. 57–58.

17. Joseph Schumpeter, *Capitalism, Socialism and Democracy* (New York: Routledge, 1943).

18. As chairman of the FCC, I tried to implement rudimentarily agile policies in three areas—net neutrality, network privacy, and cybersecurity. All three were subsequently repealed during the Trump administration in favor of no regulation.

19. Wheeler, Verveer, and Kimmelman, "New Digital Realities."

CHAPTER 10

1. 18 USC Section 1702 provides fines and up to five years in prison if a person designs "to pry into the business or secrets of another," 18 U.S. Code, sec. 1702, https://www.law.cornell.edu/uscode/text/18/1702.

2. Heather Kelly and Emily Guskin, "Americans Widely Distrust Facebook, TikTok and Instagram with Their Data, Poll Finds," *Washington Post*, December 22, 2021, https://www.washingtonpost.com/technology/2021/12/22/tech-trust-survey/.

3. Bree Fowler, "Americans Want More Say in the Privacy of Personal Data," *Consumer Reports*, May 18, 2017, https://www.consumerreports.org/privacy/americans-want-more-say-in-privacy-of-personal-data-a5880786028/.

4. Joe Mandese, "For Most Americans, Personal Data Privacy Now Rivals the Bill of Rights," MediaPost, September 12, 2018, https://www.mediapost.com/publications/article/324980/for-most-americans-personal-data-privacy-now-riva.htm.

5. https://www.law.cornell.edu/cfr/text/47/part-64/subpart-U.

6. For the FCC Customer Proprietary Network Information Rule, see the website at https://www.ecfr.gov/current/title-47/chapter-I/subchapter-B/part-64/subpart-U.

7. Wheeler quoted in Cecilia Kang, "Broadband Providers Will Need Permission to Collect Private Data," *New York Times*, October 27, 2016, https://www.nytimes.com /2016/10/28/technology/fcc-tightens-privacy-rules-for-broadband-providers.html.

8. Cecilia Kang, "Congress Moves to Strike Internet Privacy Rules from Obama Era," *New York Times*, March 23, 2017, https://www.nytimes.com/2017/03/23/technology/ congress-moves-to-strike-internet-privacy-rules-from-obama-era.html.

9. Ibid.

10. Cecilia Kang, "Congress Moves to Overturn Obama-Era Online Privacy Rules," *New York Times*, March 28, 2017, https://www.nytimes.com/2017/03/28/technology/ congress-votes-to-overturn-obama-era-online-privacy-rules.html.

11. European Commission, "Data Protection in the EU," https://commission.europa .eu/law/law-topic/data-protection/data-protection-eu_en.

12. Quoted in Marko Saric, "How To Limit Your Exposure to the Surveil- lance Capitalism," Medium, May 30, 2019, https://medium.com/swlh/surveillance -capitalism-9294fc3a7709.

13. Sergey Brin and Lawrence Page, "The Anatomy of a Large-Scale Hypertextual Web Search Engine," Computer Science Department, Stanford University (April 1998) http://ilpubs.stanford.edu:8090/361/.

14. "Google Receives $25 Million in Equity Funding," *News from Google*, June 7, 1999, http://googlepress.blogspot.com/1999/06/google-receives-25-million-in-equity.html.

15. See the Wikipedia article at https://en.wikipedia.org/wiki/Google_Ads.

16. Google Form S-1, 2004, https://www.sec.gov/Archives/edgar/data/12887766 /000119312504073639/ds1.htm.

17. Louise Story and Miguel Helft, "Google Buys DoubleClick for $3.1 Billion," *New York Times*, April 14, 2007, https://www.nytimes.com/2007/04/14/technology /14DoubleClick.html.

18. Shoshana Zuboff, *The Age of Surveillance Capitalism: The Fight for a Human Future at the New Frontier of Power* (New York: PublicAffairs, 2019).

19. Shoshana Zuboff, "You Are the Object of a Secret Extraction Operation," *New York Times*, November 12, 2021, https://www.nytimes.com/2021/11/12/opinion/facebook -privacy.html.

20. David Nield, "All the Ways Google Tracks You—And How to Stop It," *Wired*, May 27, 2019, https://www.wired.com/story/google-tracks-you-privacy/.

21. Faizah Imani, "Can You Be Tracked on YouTube?," Small Business—Chron, https://smallbusiness.chron.com/can-tracked-youtube-32762.html.

22. Counterpoint Research, "iOS vs Android Quarterly Market Share," May 15, 2023, https://www.counterpointresearch.com/gobal-smartphone-os-market-share/.

23. CompaniesMarketCap, *Largest Companies by Market Cap* (database), https:// companiesmarketcap.com/.

24. Sara Lebow, "Digital Will Account for 71.8% of US Media Ad Spend This Year, Up 16 Percentage Points from 2018—and Growing," *Insider Intelligence*, November 7, 2022, https://www.insiderintelligence.com/content/digital-us-media-ad-spend.

25. Statista, "Share of Amazon, Facebook, and Google in net digital ad revenue in the United States from 2019 to 2023," https://www.statista.com/statistics /242549/digital-ad-market-share-of-major-ad-selling-companies-in-the-us-by-revenue /#:~:text=Digital%20ad%20revenue%20share%20in%20the%20U.S.%202019%2D2023 %2C%20by%20company&text=In%202021%2C%20Google%20accounted%20for,23.8 %20and%2011.3%20percent%2C%20respectively.

26. "Facebook's Mark Zuckerberg Says Sorry to Britons with Newspaper Apology Ads," *The Wire*, March 25, 2018, https://thewire.in/world/mark-zuckerberg-cambridge -analytica.

27. Nicholas Confessore, "Cambridge Analytica and Facebook: The Scandal and The Fallout So Far," *New York Times*, April 4, 2018, https://www.nytimes.com/2018/04/04/us /politics/cambridge-analytica-scandal-fallout.html.

28. Bobbie Johnson, "Privacy No Longer a Social Norm, Says Facebook Founder." *Guardian*, January 11, 2010, https://www.theguardian.com/technology/2010/jan/11/ facebook-privacy.

29. Meta, "An Open Conversation with Rochelle," YouTube (video), October 13, 2021, https://www.youtube.com/watch?v=qV79Dvwqo9Y.

30. Heather Kelly, Chris Velazzco, and Tatum Hunter, "Amazon's Newest Products Expand Its Surveillance Inside the Home," *Washington Post*, September 28, 2021, https://www.washingtonpost.com/technology/2021/09/28/amazon-event-echo-ring -launch/.

31. David B. Guralnik, ed., *Webster's New World Dictionary of the American Language, Second College Edition* (New York: Simon & Schuster, 1980), p. 1131.

32. U.S. Government, Department of Health and Human Services, *Records, Computers and the Rights of Citizens*, Report of the Secretary's Advisory Committee on Automated Personal Data Systems, June 30, 1973, https://aspe.hhs.gov/reports/records-computers -rights-citizens.

33. Ibid.

34. Woodrow Hartzog, "The Inadequate, Invaluable Fair Information Practices," *Maryland Law Review* 76 (2017): 952, https://digitalcommons.law.umaryland.edu/mlr /vol76/iss4/4/.

35. Zuboff, "You Are the Object of a Secret Extraction Operation."

36. U.S. Government, Department of Health and Human Services, *Records, Computers, and the Rights of Citizens*.

37. "Adhesion Contract (Contract of Adhesion)," Cornell Law School, https://www .law.cornell.edu/wex/adhesion_contract_(contract_of_adhesion).

38. Lisa M. Austin, "Enough about Me: Why Privacy Is about Power, Not Consent (or Harm)," in *A World without Privacy: What Law Can and Should Do?*, edited by Austin Sarat (Cambridge: Cambridge University Press, 2014), chap. 3, https://www.cambridge .org/core/books/abs/world-without-privacy/enough-about-me-why-privacy-is-about -power-not-consent-or-harm/BC0D2D0718C26E75630548ED64B0E6C9.

39. Johnson, "Privacy No Longer a Social Norm."

40. Azeem Azhar, *The Exponential Age: How Accelerating Technology Is Transforming Business, Politics, and Society* (New York: Diversion Books, 2021), p. 221.

41. Ann Cavoukian, "Privacy by Design: The 7 Foundational Principles," Information and Privacy Commissioner of Ontario, Canada, https://www.ipc.on.ca/wp-content/uploads/resources/7foundationalprinciples.pdf.

42. Ibid.

43. Ibid.

44. General Data Protection Regulation: Data Protection by Design and by Default, Intersoft Consulting, Art. 25, GDPR, https://gdpr-info.eu/art-25-gdpr/.

45. Clare Naden, "Data Privacy by Design: A New Standard Ensures Consumer Privacy at Every Step," ISO, May, 11, 2018, https://www.iso.org/news/ref229.html.

46. Federal Trade Commission, "FTC Issues Final Commission Report on Protecting Consumer Privacy: Agency Calls on Companies to Adopt Best Privacy Practices," FTC, March 26, 2012, https://www.ftc .gov/news-events/press-releases/2012/03/ftc-issues-final-commission-report-protecting-consumer-privacy.

47. Awanthika Senarath and Nalin A. G. Arachchilage, "Why Developers Cannot Embed Privacy into Software Systems? An Empirical Investigation," in ACM, *Ease 18: Proceedings of the 22nd International Conference on Evaluation and Assessment in Software Engineering*, June 28, 2018, pp. 211–216, https://doi.org/10.1145/3210459.3210484.

48. Kathrin Bednar, Sarah Spiekermann, and Marc Langheinrich, "Engineering Privacy by Design: Are Engineers Ready to Live Up to the Challenge?," *Information Society* 35, no. 3 (2019), https://www.tandfonline.com/doi/full/10.1080/01972243.2019.1583296.

49. Connectivity Standards Alliance, "The CHIP," May 11, 2021, https://csa-iot.org/newsroom-chip-is-now-matter/.

50. Ibid.

51. Matter—SmartHome (Germany), "Benefits of Matter #4: Security and Privacy," FAQs," https://matter-smarthome.de/en/benefits/#:~:text=Benefit%20%234%20%E2%80%93%20Security%20and%20privacy,a%20high%20level%20of%20security.

CHAPTER 11

1. *Standard Oil Co. of New Jersey v. United States*, 221 U.S. 1 (1911), https://supreme .justia.com/cases/federal/us/221/1/.

2. Amy Klobuchar, *Antitrust: Taking on Monopoly Power from the Gilded Age to the Digital Age* (New York: Alfred A. Knopf, 2021), p. 73.

3. Berfield, *The Hour of Fate*, p. 49.

4. Klobuchar, *Antitrust*, p. 63.

5. 15 U.S.C. §§ 1–38, https://www.law.cornell.edu/uscode/text/15/chapter-1.

6. U.S. Department of Justice, "Competition and Monopoly: Single-Firm Conduct Under Section 2 of the Sherman Act, Chapter 2," updated on February 16, 2022, https://www.justice.gov/atr/competition-and-monopoly-single-firm-conduct-under-section-2-sherman-act-chapter-2.

7. Berfield, *The Hour of Fate*, p. 81.

8. Klobuchar, *Antitrust*, p. 41.

9. Ibid.

10. Klobuchar, *Antitrust*, p. 88.

11. Bhu Srinivasan, *Americana: A 400-Year History of American Capitalism* (New York: Penguin Press, 2017), p. 209.

12. Ibid., p. 210.

13. Doris Kearns Goodwin, *The Bully Pulpit: Theodore Roosevelt, William Howard Taft, and the Golden Age of Journalism* (New York: Simon & Schuster, 2013), p. 11.

14. *The Standard Oil Co. of New Jersey, et al. v. The United States*, 221 U.S. 1 (1911).

15. Goodwin, *The Bully Pulpit*, p. 300.

16. Ibid.

17. George E. Mowry, *The Era of Theodore Roosevelt* (New York: Harper Bros., 1958), p. 133.

18. Klobuchar, *Antitrust*, p. 104.

19. Theodore Roosevelt, "Seventh Annual Message to Congress," December 3, 1907, in *Presidential Speeches*, University of Virginia, Miller Center, https://millercenter.org/the-presidency/presidential-speeches/december-3-1907-seventh-annual-message.

20. Ibid.

21. Tim Wu, *The Curse of Bigness: Antitrust in the New Gilded Age* (New York: Columbia Global Reports, 2018), p. 66.

22. Roosevelt, "Seventh Annual Message."

23. Jonathan Sallet, "Louis Brandeis: A Man for This Season," *Colorado Technology Law Journal* (March 2018): 366–67, http://ctlj.colorado.edu/wp-content/uploads/2018/09/5-Sallet-8.4.18-FINAL.pdf.

24. L. D. Brandeis and C. M. Lewis, *The Curse of Bigness: Miscellaneous Papers of Louis D. Brandeis* (New York: Viking Press, 1934).

25. Sallet, "Louis Brandeis."

26. Sallet, "Louis Brandeis," p. 367.

27. FTC Act Section 5(c)15 U.S.C. § 45.

28. *United States v. Grinnell Corp.*, 384 U.S. 563 (1966).

29. Section 2 of the Sherman Act males it illegal to "monopolize, or attempt to monopolize, or combine, or conspire . . . to monopolize any part of . . . trade or commerce." (15 U.S.C. §2)

30. Section 7 of the Clayton Antitrust Act forbids acquisitions "the effect of [which] may be substantially to lessen competition, or to tend to create a monopoly" (15 U.S.C. § 7).

31. Herbert Hovenkamp, "The Monopolization Offense," *Ohio State Law Journal* 61 (2000): 1035, 1049, https://kb.osu.edu/bitstream/handle/1811/70413/OSLJ_V61N3_1035.pdf?sequence=1&isAllowed=y.

32. Robert Bork, *The Antitrust Paradox* (New York: Basic Books, 1978).

33. Makan Delrahim, "And Justice for All: Antitrust Enforcement and Digital Gatekeepers," remarks at Antitrust New Frontiers Conference, Tel Aviv, Israel, June 11, 2019, https://www.justice.gov/opa/speech/assistant-attorney-general-makan-delrahim-delivers-remarks-antitrust-new-frontiers.

34. Philip Verveer, "Platform Accountability and Contemporary Competition Law: Practical Considerations," Discussion Paper, Shorenstein Center, Harvard

Kennedy School, November 2018, https://shorensteincenter.org/platform-accountability
-contemporary-competition-law-practical-considerations/.

35. Sallet, "Louis Brandeis," p. 366.

36. *Standard Oil Co. of New Jersey v. United States*, 221 U.S. 1 (1911).

37. *United States v. American Telephone & Telegraph Co.*, 552 F. Supp. 131 (1983).

38. Harold Feld, *The Case for the Digital Platform Act: Market Structure and Regulation of Digital Platforms*, May, 8 2019, available online from Roosevelt Institute, https://rooseveltinstitute.org/publications/the-case-for-the-digital-platform-act-market
-structure-and-regulation-of-digital-platforms/.

39. *United States v. American Telephone & Telegraph Co.* began in 1973 and the ordered divestitures went into effect in 1984. *United States v. Microsoft Corp.* took even longer, starting with an FTC investigation in 1990 and ending with appellate court approval of a settlement in 2004. See Philip Verveer, *Platform Accountability.*

40. George Stigler Center for the Study of the Economy and the State, The University of Chicago Booth School of Business Committee on the Study of Digital Platforms, Market Structure and Antitrust Subcommittee Report 21, July 2019, https://www
.chicagobooth.edu/research/stigler/news-and-media/committee-on-digital-platforms
-final-report.

41. "Best Technological Inventions in Last Decade That Changed Technology Sector," Your Tech Diet, https://yourtechdiet.com/blogs/10-best-inventions-last-decade
-technology/.

42. Wheeler, Verveer, and Kimmelman, New Digital Realities, p. 33.

43. Carl Shapiro, "Protecting Competition in the American Economy: Merger Control, Tech Titans, Labor Markets," *Journal of Economic Perspectives*, vol. 33, no. 3 (Summer 2019), pp. 69–96, https://www.aeaweb.org/articles?id=10.1257/jep.33.3.69.

44. Roosevelt, "Seventh Annual Message to Congress."

45. Ibid.

46. Ibid.

47. Fiona Scott Morton et al., "Equitable Interoperability: The 'Super Tool' of Digital Platform Governance," Policy Discussion paper No. 4, Yale Tobin Center for Economic Policy, July 2021, https://tobin.yale.edu/sites/default/files/Digital%20Regulation%20Project
%20Papers/Digital%20Regulation%20Project%20-%20Equitable%20Interoperability%20
-%20Discussion%20Paper%20No%204.pdf.

48. Ibid., p. 2.

49. Posting a YouTube link on Facebook is not interoperability; rather, it forces the user to link out to the interoperability of the internet instead of displaying the video on the Facebook feed.

50. Bill Roberts, "Celebrating the First Anniversary of Open Banking," Competition and Markets Authority, Jan. 11, 2019, https://competitionandmarkets.blog.gov.uk/2019
/01/11/open-banking-anniversary/.

51. Ibid.

52. Author's discussion with Imran Gulamhuseinwala, former trustee, Open Banking, Ltd., February 2022.

53. Ibid.

54. Mark Zuckerberg, "Four Ideas to Regulate the Internet," *Washington Post*, June 10, 2021, https://www.washingtonpost.com/opinions/mark-zuckerberg-the-internet -needs-new-rules-lets-start-in-these-four-areas/2019/03/29/9e6f0504-521a-11e9-a3f7 -78b7525a8d5f_story.html.

55. Data Transfer Project, July 20, 2018, https://datatransferproject.dev/.

56. Rene Kolga and Nelly Porter, "Introducing Confidential Space to Help Unlock the Value of Secure Data Collaboration," October 1, 2022, https://cloud.google.com/blog/ products/identity-security/announcing-confidential-space.

57. Amazon, "AWS Announces AWS Clean Rooms," November 29, 2022, https:// press.aboutamazon.com/2022/11/aws-announces-aws-clean-rooms.

58. Kogla and Porter, "Introducing Confidential Space."

59. Amazon, "AWS Announces AWS Clean Rooms."

Chapter 12

1. Eric Burns, *Infamous Scribblers: The Founding Fathers and the Rowdy Beginnings of American Journalism* (New York: PublicAffairs Press, 2007), p. 3.

2. Ibid., p. 12.

3. Calvin Yu, "Printing History: Steam Powered Printing," *The Spill* (blog), Ink Toner, June 3, 2013, http://www.247inktoner.com/blog/post/2013/06/03/Printing -History-Steam-Powered-Printing.aspx.

4. Paul Starr, *The Creation of the Media: Political Origins of Modern Communications* (New York: Basic Books, 2004), p. 133.

5. Gerald Chait, "Half the Money I Spend on Advertising Is Wasted; The Trouble Is I Don't Know Which Half," B2B Marketing, March 18, 2015, https:// www.b2bmarketing.net/archive/half-the-money-i-spend-on-advertising-is-wasted-the -trouble-kis-i-dont-know-which-half-b2b-marketing/.

6. Quoted in Porter Bibb, *It Ain't As Easy As It Looks: Ted Turner's Amazing Story* (New York: Crown, 1993), p. 180.

7. Randall S. Sumpter, *Before Journalism Schools: How Gilded Age Reporters Learned the Rules* (Columbia: University of Missouri Press, 2018), p. 94.

8. Jonathan Rauch, *The Constitution of Knowledge: A Defense of Truth* (Washington, DC: Brookings Institution Press, 2021), p. 122.

9. Frank Luther Mott, *American Journalism: A History, 1690–1960* (New York: Macmillan, 1962). See also https://en.wikipedia.org/wiki/WilliamRandolphHearst.

10. Ibid. See also https://en.wikipedia.org/wiki/JosephPulitzer.

11. "Yellow Journalism: The 'Fake News' of the 19th Century," *Public Domain Review*, https://publicdomainreview.org/collection/yellow-journalism-the-fake-news-of -the-19th-century or https://history.state.gov/milestones/1866-1898/yellow-journalism.

12. Frank Luther Mott, *American Journalism: A History, 1690–1964* (New York: Macmillan, 1962), p. 539.

13. Ibid.

14. American Society of Newspaper Editors, "Statement of Principles," 1923, https:// accountablejournalism.org/ethics-codes/american-society-of-newspaper-editors -statement-of-principles.

15. Rauch, *The Constitution of Knowledge*, p. 126.

16. Mattha Busby, "Social Media Copies Gambling Methods 'to Create Psychological Cravings,'" *Guardian*, May 8, 2018, https://www.theguardian.com/technology/2018/may/08/social-media-copies-gambling-methods-to-create-psychological-cravings.

17. Roger McNamee, *Zucked: Waking Up to the Facebook Catastrophe* (New York: Penguin Press, 2019), p. 84.

18. Sam Harris, *Making Sense* podcast #148, February 5, 2019, https://www.samharris.org/podcasts/making-sense-episodes/148-jack-dorsey.

19. *Grosjean v. American Press Co., Inc.*, 297 U.S. 233 (1936), https://supreme.justia.com/cases/federal/us/297/233/#246.

20. The fairness doctrine appeared in "Report on Editorializing by Broadcast Licensees," 13 F.C.C.1246 (1949), later amended to the Communications Act of 1934, 47 U.S.C.

21. *Red Lion Broadcasting Co., Inc. v. Federal Communications Commission*, 395 U.S. 367 (1969), https://supreme.justia.com/cases/federal/us/395/367/#390.

22. The fairness doctrine was subsequently repealed by the Reagan FCC in 1987.

23. The fairness doctrine was ultimately dismantled by the Reagan FCC. When Congress passed a law reinstating it, President Reagan vetoed the bill.

24. Code of Practices for Television Broadcasters adopted December 6, 1951, by the National Association of Broadcasters, http://www.tvhistory.tv/SEAL-Good-Practice1.JPG.

25. In 1976 the code's requirement for a Family Viewing Hour (a suggestion from the FCC chairman) was overturned by a court decision as an unconstitutional effort of a government agency. In 1979 the Justice Department alleged the agreement to limit the number of advertisements was a Sherman Act violation, and the NAB settled by dropping the requirement. While these experiences exemplify the challenge of code creation, they should not, however, discourage the search for solutions.

26. Mark MacCarthy, *Regulating Digital Industries: How Public Oversight Can Encourage Competition, Protect Privacy, and Ensure Free Speech* (Washington, DC: Brookings Institution Press, 2023).

27. This idea was first proposed by Wael Ghonim, the Google employee who organized the Tahrir Square uprising in Egypt in 2011 while we were together at the Harvard Kennedy School. See Tom Wheeler, "How to Monitor Fake News," *New York Times*, February 20, 2018, https://www.nytimes.com/2018/02/20/opinion/monitor-fake-news.html.

28. Abraham Lincoln, "Address before the Young Men's Lyceum of Springfield, Illinois," January 27, 1838, http://www.abrahamlincolnonline.org/lincoln/speeches/lyceum.htm.

29. Abraham Lincoln, Annual Message to Congress, December 1, 1862, https://www.abrahamlincolnonline.org/lincoln/speeches/congress.htm.

Consequences We Control

1. Edward O. Wilson, Debate at Harvard Museum of Natural History, Cambridge, Massachusetts, September 9, 2009, https://www.oxfordreference.com/view/10.1093/acref/9780191826719.001.0001/q-oro-ed4-00016553.

CHAPTER 13

1. Matt Hamblen, "FCC a 'Referee,' Not a Regulator, of the Internet, Wheeler Says," *Computerworld*, March 3, 2015, https://www.computerworld.com/article/2892317/fcc-a -referee-not-a-regulator-of-the-internet-wheeler-says.html.

2. J. David Hacker, "Decennial Life Tables for White Population of the United States, 1790–1900," National Library of Medicine, https://www.ncbi.nlm.nih.gov/pmc/articles /PMC2885717/.

3. Simon Van Dorpe and Leah Nylen, "Europe Failed to Tame Google. Can the U.S. Do Any Better?," Politico, October 21, 2020, https://www.politico.com/news/2020/10/21/ google-europe-us-antitrust-431036.

4. Meta Platforms, Inc., 10-K, https://investor.fb.com/financals/default.aspx.

5. Eric Schmidt and Jared Cohen, *The New Digital Age: Transforming Nations, Businesses, and Our Lives* (Knopf Doubleday Publishing Group, 2013), pp. 9, 10.

6. Gus Lubin, "Peter Thiel Is Trying to Save the World: The Apocalyptic Theory behind His Actions," *Business Insider*, December 8, 2016, https://www.businessinsider .com/peter-thiel-is-trying-to-save-the-world-2016-12.

Index

219

About the Author

Businessman, venture capitalist, and former chairman of the Federal Communications Commission during the Obama administration, **Tom Wheeler** is the author of several books, including, most recently, *From Gutenberg to Google: The History of Our Future* (Brookings, 2019). He resides in Washington, D.C.